The Call of Coincidence

Mathematical Gems, Peculiar Patterns, and More Stories of Numerical Serendipity

OWEN O'SHEA

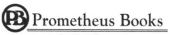

Prometheus Books

Guilford, Connecticut

 Prometheus Books

An imprint of Globe Pequot, the trade division of The Rowman & Littlefield Publishing Group, Inc.
4501 Forbes Blvd., Ste. 200
Lanham, MD 20706
www.rowman.com

Distributed by NATIONAL BOOK NETWORK

British Library Cataloguing in Publication Information Available

Library of Congress Cataloging-in-Publication Data Available

Names: O'Shea, Owen, 1956– author.
Title: The call of coincidence: mathematical gems, peculiar patterns, and more stories of numerical serendipity/Owen O'Shea.
Description: Lanham, MD: Prometheus, 2023. | Includes bibliographical references. | Summary: "Featuring surprising trivia gems alongside serious questions like why there is something rather than nothing, readers will be enriched by this exploration of remarkable number coincidences and the mathematics that make them possible—and probable"—Provided by publisher.
Identifiers: LCCN 2022056222 (print) | LCCN 2022056223 (ebook) | ISBN 9781633889262 (paperback) | ISBN 9781633889279 (epub)
Subjects: LCSH: Mathematical recreations. | Probabilities—Problems, exercises, etc.
Classification: LCC QA95 .O74 2023 (print) | LCC QA95 (ebook) | DDC 793.74—dc23/eng20230415
LC record available at https://lccn.loc.gov/2022056222
LC ebook record available at https://lccn.loc.gov/2022056223

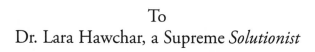

To
Dr. Lara Hawchar, a Supreme *Solutionist*

Contents

Acknowledgments

I wish to thank Layal Hawchar for the excellent drawings created specifically for this work.

I wish to thank the Sam Loyd Company for its kind permission to reproduce the drawing accompanying the Tom the Piper's Son puzzle, reproduced from Sam Loyd's *Cyclopedia of Puzzles*.

The name "Sam Loyd" is a registered trademark of the Sam Loyd Company.

Introduction

Recreational mathematics is an umbrella term that is used to describe a vast potpourri of mathematics that recreational mathematicians, most of whom are amateur, pursue for intellectual fun and pleasure. This volume contains chapters on several topics in recreational mathematics. Some of these topics have been discussed elsewhere in previous works; however, I have tried to include the most up-to-date information for the reader on the various topics discussed.

The opening chapter concerns the theory of probability and the phenomena of strange coincidences (with several interesting examples) and why they are inevitable in the world we live in.

In this book, you will find a discussion on Waring's problem and how a beautiful, simple mathematical formula appears to give the correct answer to an infinite number of questions that arise in Waring's problem. The only downside with the formula is that mathematicians have still not proved that it holds in *all* cases.

Recreational mathematics enthusiasts are a curious breed. I have therefore included an extensive chapter that concerns the ancient question on whether mathematics is *invented* or *discovered.*

There are discussions concerning the theory of *simplicity* in mathematics and *simplicity* in nature and whether the universe we live in operates according to simple laws, complex ones, or both. The unanswerable metaphysics question of *why there is something rather than nothing* is also discussed. Deep-thinking readers will not want to miss the discussion of this staggering question.

There is a chapter devoted to explaining (with examples) how mathematicians derive formulas to solve problems in mathematics.

There are also chapters on less serious topics to maintain a balance in the book's appeal to a wide readership.

For example, I have included three chapters giving various details conveyed to me by fictional numerologist Dr. Moogle, who never fails to surprise and delight. Dr. Moogle is an extraordinary man who I am so fortunate to know. He has an awesome knowledge of the properties of numbers. His beautiful daughter, Anna, is also extraordinarily gifted at discovering interesting connections between numbers. Dr. Moogle has given me many brilliant number coincidences. He has also given me completely original parallels in the lives of outstanding American puzzle creator Sam Loyd and the great popularizer of mathematical recreations Martin Gardner.

Recreational mathematics enthusiasts around the world, as well as soccer fans all over the globe, will particularly delight in two chapters in this volume. In one memorable chapter, Dr. Moogle gives original, never-before-published, remarkable number coincidences concerning the 1966 FIFA World Cup in England.

Another fictional character I met on my recent travels is Jack O'Mara, known to all and sundry as "Mara." He is a hardworking, intelligent, and inquisitive seaman, known to seafarers all over the world as being quite accomplished at spotting original coincidences involving numbers.

One of the most gifted fictional men at discovering and unearthing number coincidences is undoubtedly Charlie Chance. I knew him when he worked as a bartender in Dublin. Charlie left Ireland's capital about 20 years ago, when he was attracted by the sand and sea of Spanish holiday resorts. He worked in the sunny land of Spain for a number of years, serving in bars and bistros, where he met all sorts from all corners of the world. Charlie contacted me in recent months. I met up with him in Manchester in the United Kingdom when I was there recently on private business. Charlie has always been interested in number coincidences, and soccer. He gave me quite remarkable coincidences concerning Manchester United, one of the most famous soccer clubs in the world.

The Call of Coincidence contains many original and interesting number coincidences that appear here in print for the first time concerning the fabulous and retired Cunard ocean liner *QE2*; the famous American ore carrier SS *Edmund Fitzgerald*, which sank in a storm in 1975; and the late and much-loved *Queen Mother* of the United Kingdom.

I believe readers with an inquiring mind and a sense of fun should consider purchasing this book. Hopefully, they will find a work that is original, different, enjoyable, and, may I say, also educational.

Owen O'Shea
Ireland
January 2023

CHAPTER 1

Probability and Coincidence

Strange coincidences have fascinated people down through the centuries. Many of these people have probably believed that these strange occurrences are evidence that the paranormal exists. When such events happened, people immediately attributed their causes to God, the devil, or to some other occult forces.

Of course, skeptics have long argued that strange coincidences, even the most bizarre of them, can be expected to occur from time to time. Today, they argue that in a world containing nearly 8 billion people, many of whom are doing hundreds of different things each day, there are bound to be some events that we human beings find startling.

There are numerous coincidences that appear in nature. Here are two of the most famous.

The sun and moon look approximately the same size when viewed from Earth. It is this apparent similarity in size that allows solar eclipses to occur. The sun's diameter is about 400 times bigger than that of the moon, but the sun is also about 400 times farther from Earth than the moon is. This coincidence makes the two objects appear the same size when viewed from Earth.

The second coincidence concerns the fact that ice floats on water. Most liquids, when frozen and immersed in their own liquid, sink. Water is virtually unique in that this does not apply to it. When water freezes, the distance between its molecules increases very slightly. This makes water in its frozen state less dense than when it is in its liquid state, thereby allowing the ice to float. If water did not have this property, no living things could exist on this planet. The various ice ages would have meant that the oceans would have become full of ice, killing all life forms in them. It is a sobering thought that you and I (and all other living things on Earth) are only here because ice floats in water.

Some people are naturally skeptical of the occurrence of coincidences. They just do not see any reason to be excited by the occurrence of a strange, surprising, and unusual event. One of the most skeptical was the famous American physicist, Richard Feynman (1920–1988), who once said, "The most amazing thing happened to me tonight. I saw a car with the license plate ARW 357. Can you imagine? Of all the millions of license plates in the state, what was the chance that I would see that particular one tonight?"

Of course, Feynman was correct in what he had said. The license plate he saw was undoubtedly one of millions that he could have seen but didn't. Feynman was making fun of people who are genuinely surprised by strange *and* surprising coincidences. The point Feynman was making was that he had to see *some* unique license plate on the night in question and that a rare event such as this happens all the time. No one denies this. It was a rare license plate, but it was not a *surprising* one. That is the difference—and it is a big difference—between what he saw and what people experience when they encounter a genuine coincidence. Suppose Feynman had seen a license plate that read "ABC 123"? In that case, the license plate number is also unique, but it is also ordered according to an intelligent and ordered pattern that is identifiable and understood by the average human mind. It is this difference that creates a pleasant surprise to the observer. The human mind would perceive the successive order of the first three consecutive letters and the first three consecutive digits of the license plate and realize that this is rather *unique* and *surprising.*

Thus, to define coincidences as merely *unlikely events* is to ignore a key component of coincidences: their *apparent meaningfulness.* A coincidence may be defined as an accidental occurrence of two events that strongly but *falsely* suggests a causal relationship between the two events.

Saying the same thing another way, a formal definition of coincidences is that they are events that provide support for a hypothesis that one ultimately decides is false. Thus, when we encounter a remarkable coincidence, it pleasantly stuns us, because it makes us temporarily question our belief on how the world operates.

Suppose for, example, that I have five identical cards, each bearing a number from 1 to 5 on one side of the card. The other side of each card is blank. I ask you to thoroughly shuffle the five cards and then ask you to place the five cards face down on a nearby table. I point out to you that the five cards can be arranged in 120 different ways. I also point out that the arrangement 1, 2, 3, 4, 5 is just one of those 120 ways. Hence, the probability that the cards are in that specific order is 1 chance in 120. In other words, that outcome is very unlikely.

You turn over the five cards, and you see that they are arranged in the following order: 3, 2, 5, 1, 4. You are not surprised because, from your perspective, the cards appear in a random order. That is what you and everyone else expected to happen, although the probability that the cards would be arranged in the order 3, 2, 5, 1, 4 is also 1 chance in 120.

However, suppose the five cards appeared in the following order: 1, 2, 3, 4, 5. This would surprise you and everyone else witnessing it because this is not the way the world *usually* works. The human mind acknowledges that this pattern is ordered in a *meaningful* way. Hence, all who witness this event would experience a pleasant coincidence.

It is this apparent meaningfulness accompanying the rare event that constitutes a genuine coincidence.

Noticing clear coincidences can sometimes lead to scientific discovery. If Halley had not noticed the plain coincidence of a comet tracking a remarkably similar path

across the sky in 1531, 1607, and 1682 and concluding that it was not three different comets but the same comet appearing every 76 years or so, he might never have
discovered the comet named in his honor.

Strange but trivial coincidences occur much of the time in everyday life, but most
go unnoticed. For example, no one has apparently noticed the following coincidence:
Air France Flight 8969 was due to fly from Houari Boumedienne Airport, Algiers, at
11:15 a.m. (local time) on December 24, 1994, when it was hijacked. Curiously, 8,969
happens to be the 1,115th prime number.

A trivial mathematical coincidence may occur when a person is performing some
calculations. For example, one may notice that the number of feet in a mile (5,280)
equals $(10 \times 11 \times 12 \times 13) - (9 \times 10 \times 11 \times 12)$. The person noticing a coincidence
such as this may be pleasantly surprised to notice the repetition of the numbers 10,
11, and 12. But it is unlikely that she will consider the coincidence mind-blowing.

Other trivial mathematical coincidences that have been noted are that the 12th
prime is 37 and the 21st prime is 73. Note the reversals of the index number of the
prime and the prime itself. Also, the 13th digit of the three irrational numbers π (pi),
e (Euler's number), and ϕ (phi, the golden ratio) is 9:

$$\pi = 3.14159265358\underline{9}79 \ldots$$

$$e = 2.71828182845\underline{9}04 \ldots$$

$$\phi = 1.61803398874\underline{9}89 \ldots$$

The string of four digits, 1828, appear at the second and sixth decimal positions
of e. The repetition of those digits, so early in the decimal expansion of e, is purely
coincidental. The next appearance in e, of 1828, appears at the 30,569th decimal position. Also coincidental is the appearance of six consecutive nines in π, beginning at
the 762nd decimal position. The next appearance of six consecutive nines in π begins
at the 193,034th decimal position.

Trivial coincidences crop up in the lives and deaths of many notorious criminals,
including one of the most infamous of all, the American gangster Al Capone. He was
born in Brooklyn, New York, on January 17, 1899, and died in Palm Island, Mi-
ami Beach, Florida, on January 25, 1947. He lived for 17,539 days and died on the
$(1 + 7 + 5 + 3 + 9)$ day of the year.[1]

These are all insignificant coincidences, of passing interest only to those interested
in numbers. However, some coincidences are so rare that they make headlines when
one reads or hears about them.

For example, Carolyn and Ralph Cummins of Clintwood, Virginia, had five children.
Curiously, all five children were born on February 20. Catherine was born in 1952, Carol
came into the world in 1953, Charles was delivered in 1956, Claudia arrived in 1961, and
Cecillia made her entrance in 1966. The probability of this happening is 1 in more than
17 billion. The family was entered into the 1977 *Guinness Book of Records*.

Curious coincidences that caught the public's imagination have happened many
times over the years. Here are six of the more prominent.

The first one concerns the ill-fated Apollo 13 moon mission. Apollo 13 was
launched from the Kennedy Space Center at 13:13 hours on April 11, 1970. When

that date is written as 4/11/70, its digits sum to 13. The mission had to be aborted because of an explosion that occurred on board the spacecraft on April 13 (the date in Kennedy Space Center at the time of the explosion).

In 1973, actor Anthony Hopkins agreed to appear in the movie titled *The Girl from Petrovka*. The movie is based on a novel of the same name by George Feifer. While in London, Hopkins sought in vain to purchase the novel so that he would become familiar with its story line and plot. Unable to find a copy of the book in any of the bookstores in London (including secondhand stores), Hopkins headed back on the Underground (subway) at the end of a long day to his apartment, tired and disappointed. As Hopkins waited for his train, he was surprised to discover a book lying on a bench in a train station. Amazingly, the book was *The Girl from Petrovka*. But even more amazing, the book turned out to be Feifer's own copy, which Feifer had lent to a friend and which had been stolen from his friend's car.[2]

The third coincidence concerns World War I. It has been estimated that 1 million British lives were lost in that dreadful conflict. Most people are surprised to learn that the resting places of the first and last British soldiers killed in the war are laid to rest about 20 feet apart. When the two soldiers were buried, their status as the *first* and *last* British soldiers to be killed in the war was unknown.

The first British soldier killed in World War I was Private John Parr. He was killed by German fire on August 21, 1914, in the village of Obourg, near Mons, Belgium. He was born in Finchley, now a part of London, in 1897. Private Parr was the youngest of eleven children.

The last British soldier killed during World War I was Private George Ellison. He was born in Leeds in 1878. He was killed by a German sniper on November 11, 1918, just 90 minutes before the cease-fire came into effect.

By a remarkable coincidence, both soldiers are buried in the St. Symphorien Military Cemetery in Mons, about 20 feet from each other.[3]

The fourth coincidence concerns the number lottery that is held twice weekly in South Africa. To win the jackpot in the lottery, one must correctly choose five numbers from 1 to 50 plus correctly select an additional Powerball number from the numbers 1 to 20. On Tuesday, December 1, 2020, 20 players correctly chose the following five numbers: 5, 6, 7, 8, 9. They also correctly chose the Powerball number, which was 10. The draw was televised live.

The chances of doing this are 1 in 42,375,200. This probability is calculated as follows. The number of ways of choosing five numbers from 1 to 50 is $[(50 \times 49 \times 48 \times 47 \times 46)/120]$. This equals 2,118,760. There are 20 different ways of choosing the Powerball number. Thus, the number of ways of hitting the jackpot is $2,118,760 \times 20$. This equals 42,375,200. Each of the 20 people who purchased a winning ticket won $370,000.

There was an outcry on social media, with many claiming that the lottery was fixed. They claimed that there was no way the numbers 5, 6, 7, 8, 9, and 10 could be randomly drawn. These people just do not understand that very *unlikely* events are *likely* to occur from time to time. In this case, many believed that corruption had taken place. However, an internal investigation by the South African Commission concluded that the whole episode could be put down to coincidence.[4]

The fifth coincidence is actually a set of strange coincidences involving Jason Cairns-Lawrence and his partner, Jenny, both of whom lived in the English Midlands. They went on vacation to New York in September 2001. They were in the Big Apple on the 11th day of that month, when terrorists attacked New York City. Luckily, the British couple survived the attacks, which killed about 3,000 people.

Four years later, in July 2005, Jason, a sales agent with a metal plate company in Birmingham, England, and Jenny, who worked in a dental laboratory, decided to go to London for a vacation. They were in England's capital city on July 7 of that year when terrorists attacked it. Fifty-two people were killed that day, but, fortunately, Jason and Jenny survived.

Then, in November 2008, the couple went on vacation to Mumbai, India. On the 26th day of that month, terrorists attacked the city. Almost 180 people were killed in the attacks, but once again, Jason and Jenny survived.

Amazingly, the couple had been in the locations of three of the worst terrorist attacks the world has ever seen and survived all three incidents.[5]

The sixth coincidence occurred soon after the beginning of the millennium. The story was originally in the *Swindon Advertiser* of June 28, 2001. In June of that year, a young girl named Laura Buxton in Staffordshire, England, was at her grandparents' golden wedding anniversary. Laura's grandfather suggested that Laura write a message with her name and address on a tag, attach it to a helium balloon, and then release the balloon. Laura's grandfather told his granddaughter that she might gain a pen pal from whoever found the balloon.

Laura wrote the message "Please return to Laura Buxton" along with her name and address on a tag, attached it to a helium balloon, and then released the balloon, exactly as her grandfather had suggested.

Seven days later, a farmer named Andy Rivers, living in Milton Lilbourn, Wiltshire, saw a deflated balloon caught in a hedge that separated his land from that of his neighbors. He read the message on the balloon and believed it belonged to his neighbor's 10-year-old daughter, who was named Laura Buxton. Mr. Rivers at once handed over the balloon to his neighbors Peter and Eleanor Buxton, parents of Laura.

However, they soon discovered that the balloon belonged to a Laura Buxton, who lived 140 miles away. The Laura Buxton in Wiltshire wrote to the Laura Buxton in Staffordshire. With both of their parents' consent, the two Lauras met up and became good friends. They discovered that they had several things in common. One girl was 10 years old, and the other girl was just shy of 10. When they first met, both wore similar outfits. Both girls had similar eye coloring, and both had long brown hair. Both kept gray rabbits and guinea pigs, and both had three-year-old black Labradors as pets.[6]

Several notable coincidences have occurred in the history of the United States. For example, three months before Abraham Lincoln was assassinated in April 1865, his son, Robert Lincoln, was in a train station awaiting the arrival of a train. Robert fell between the train and the platform. However, he was pulled to safety by a man named Edwin Wilkes Booth. Three months later, Edwin Wilkes Booth's brother, John, assassinated Robert Lincoln's father, President Abraham Lincoln.[7]

One can get a deeper understanding of why strange events occur by considering the following. Suppose a 52-card deck of cards is shuffled, and then dealt one card at a time. One would not expect all four suits, clubs, hearts, spades, and diamonds, each suit in their correct order, to be dealt. Why? Because there are about 10^{67} different arrangements of a deck of fifty-two cards but only one way to deal the four suits in the order just stated.

Suppose, however, that there are, say, billions of decks of cards being constantly shuffled every second. In that case, one will see a deck of cards being shuffled into order not only once but an *enormous* number of times. In other words, such low-probability events will occur—indeed are highly likely to occur—because of the enormous number of decks of cards that are being constantly shuffled.

A similar principle explains why strange and rare coincidences occur. Overall, there are an enormous number of events happening throughout the world each day. The majority of these billions of events will not be noteworthy. But a tiny percentage of them will. The analogy between the shuffling of the enormous number of decks of cards and the enormous number of events occurring around the world each day should bring home to the reader that many rare and strange events are highly likely to occur often. It would be very strange if strange things did not occur.

Thus, *unspecified unlikely events* are likely to happen. When someone wins a large prize in a lottery, that person will find it incredible that they picked the winning combination of numbers. But we all know that *someone* will win eventually. This goes to the heart of why coincidences happen. A truly remarkable, *specified* coincidence is unlikely to happen to you tomorrow, but a very unlikely coincidence will happen to *someone* every day of the week.

Sometimes, a gross injustice occurs because of society's general ignorance of the laws of probability. In 1996, in England, a solicitor was found guilty of murdering her two infants. The accused had maintained that her two infants had died from sudden infant death syndrome (SIDS). She was supported throughout her case by her husband, who was also a solicitor.

The prosecution produced an expert witness who claimed that the probability of two infants dying of SIDS in one family was in the region of 1 in 73 million. The expert witness arrived at this figure by arguing that single SIDS death in a family was 1 chance in 8,543; thus, two such deaths in the one family equals 1 chance in $8,543 \times 8,543$, or approximately 1 chance in 73 million. In 2001, the accused was convicted of murder. Following her conviction, she was widely condemned in British newspapers. She was eventually released after two appeals to the courts. However, she found life extremely difficult with all that had happened to her. She was found dead in her home in Essex in 2007.[8]

The mathematical calculation used in this case by the expert witness is incorrect. Let's assume, as the expert witness did, that the given probability of one case of SIDS in a low-risk family is 1 chance in 8,543. Then the probability of two cases of SIDS in the one family being equal to 1 chance in $8,543^2$ is false.

To see this, consider the following. Suppose someone won the Powerball lottery by beating odds of 1 chance in 292,201,338. If that player bought a ticket in the very next Powerball game, her chance of winning the jackpot is still 1 chance in 292,201,338, even if she plays the exact same numbers that previously won the jackpot for her. It is incorrect to state that her chance of winning the jackpot is 1 chance in $292,201,338 \times 292,201,338$. But this is the kind of reasoning that was used by the prosecution witness in this case.

In 2001, the Royal Statistical Society issued a statement expressing its concern at the "misuse of statistics in the courts." The statement went on to say that there was no statistical basis for the figure of 1 in 73 million. In 2002, the society wrote to the Lord Chancellor saying that "the calculation leading to the 1 chance in 73 million was false." Professor of mathematics Ray Hill of the University of Salford, England, said in 2016, in an article in the journal *Medicine, Science, and the Law*, that the calculation used by the prosecuting witness in this case led to a deduction that was clear nonsense.[9]

This all demonstrates that the use of statistics in court cases can be very dangerous and misleading unless there are expert statisticians there to explain the probabilities involved.

But let's return to the fun side of coincidences. A person may vacate abroad and unexpectedly meet a neighbor who lives nearby. The person will remember this strange incident for the rest of their life. But they will not recall the numerous times they vacationed abroad and never met anyone they know. Thus, there is a specific type of bias showed when people remember coincidences.

The fact that plain order happens amid plain randomness is illustrated by considering the decimal digits in the number pi. This is a famous number in mathematics. Pi equals 3.14159265358979323. . . . Its decimal expansion goes on forever. A Google employee named Emma Iwao calculated pi to 31,415,926,535,897 decimal places (note the digits) in 2019.[10] Starting at the 60th decimal (exceedingly early in its decimal expansion), the following sequence appears: 4592307816. That sequence has all 10 digits though obviously not in order.

Do the 10 digits in order appear in the expansion of pi? Yes, they do, but it occurs first at decimal position 17,387,594,880. The next appearance of that 10-digit sequence appears at decimal position 26,852,899,245.

Mathematicians believe that pi is a *normal* number. However, they have not yet been able to prove this. If pi is normal, then every sequence of digits appears somewhere in the endless decimal expansion of pi. For example, the sequence of digits making up your birthday is in pi. The birthdays of any famous person you can think of are in pi. Albert Einstein, for example, was born on March 14, 1879. His birthday, 3141879 (note the first three digits), first appears in pi at decimal position 1,117,773.

Thus, as we examine the digits of pi and go out to billions of decimal places, we should *expect* to see several startling but limited patterns. We cannot say in advance what these patterns will be. The most we can say is that there will be *unspecified, limited*, and *unexpected* patterns as one travels out through the never-ending digits of pi.

The appearance of striking patterns in the midst of plain randomness in pi corresponds to events that happen in life. Most events that occur around the world every day are insignificant. But a tiny percentage of them proves noteworthy. Among these are the strange and delightful events we call coincidences.

The multiverse (if it exists) is governed by the same principle. The vast majority of universes are such that they cannot support life as we know it. But a tiny percentage of these universes is exactly right to allow life to begin and evolve. Obviously, if intelligent life evolves in one of those universes, that life will find itself in a universe where the parameters are set that allow life to exist.

Finally, it is reasonable to believe that it would be a sad world, indeed a strange world, if there were no coincidences. If that were the case, it would be the biggest coincidence of all.

CHAPTER 2

Cubic Numbers

One of my previous books, *The Call of the Primes* (2016), contains chapters on both the triangular numbers and the square numbers, which are two well-known number series in mathematics. A third series of numbers is that of the cubic integers. The positive cubic integers are 1, 8, 27, 64, 125, 216, 343, 512, 729, 1,000, 1,331, 1,728, and so on to infinity. An integer, n^3, is called a cubic integer because it is the volume of a cube that has a side length of n. In other words, $n \times n \times n = n^3$.

I will try to include in this chapter cubic number properties that may be unknown to many recreational mathematics enthusiasts.

First, it should be noted that solutions to problems involving natural phenomena often include cubic equations. We shall, for the sake of completeness, briefly discuss some of these here.

The time for evaporation of a black hole due to Hawking radiation is proportional to the cube of its mass.[1]

The tidal effect produced on Earth by any body in space is proportionate to the mass of that body, but that tidal effect decreases as the *cube* of its distance. For example, consider the relative difference between the tidal pull of the sun and the moon on Earth. The sun is about 27 million times as massive as the moon. The sun is also about 390 times farther from Earth than the moon is. If we divide 27,000,000 by the cube of 390 (59,319,000), we obtain 0.455166. . . . Thus, the tidal effect of the sun is about 0.46 times that of the moon.[2]

Consider a camper sitting, say, 1 unit in distance from a campfire. The heat from the campfire is determined by the amount of wood in the fire. If the camper moves 2 units from the campfire, the amount of heat she will receive from the fire is 2^3, or 8, times less than if she were 1 unit in distance from the campfire. Of course, if she insists on sitting 2 units from the fire, she can still receive the same amount of heat as if she were 1 unit from the fire but only if 8 times as much wood is placed in the fire.[3]

Suppose you are rowing a boat in the water. To double the speed of the boat, you need to supply 2^3, or 8, times more power. If you just double the amount of power, the boat's speed increases at the rate of the cube root of 2, or 1.25992.[4]

Or consider wind turbines. These are usually placed in windy areas. There is a logical reason for this. The wind power of a wind turbine increases with the *cube* of the wind speed. Thus, if the wind speed doubles, it will result in 2^3, or 8, times the wind power.[5]

Cubic quantities appear in *Kepler's third law of planetary motion*. Consider the solar system, which consists of the sun and the eight planets orbiting it. The average distance of Earth from the sun is about 93 million miles. Let this distance equal one *astronomical unit*. We know that Earth orbits the sun once every year. Suppose we want to find out how long it takes Jupiter to orbit the sun. We look up a table and find that Jupiter is at an average distance of 484 million miles from the sun. This distance equals about 5.2043 astronomical units. If we cube 5.2043, we obtain 140.9571045. Obtain the square root of this. The result is about 11.87253 years. This is the length of time it takes Jupiter to orbit the sun. The same rule can be used to determine the orbital periods of the other seven planets as they orbit the sun.[6]

From a recreational mathematician's point of view, cubic numbers are remarkably interesting. Bui Quang Tuan noticed that if one writes the integers in the following array (see the illustration on page 11), the sum of all the integers on the left-hand side equals a cubic number.[7]

Every multiple of 6 can be expressed as the sum of four signed cubes. This is because of the following algebraic identity:

$$6x = (x + 1)^3 + (x - 1)^3 - (x^3) - (x^3)$$

Several problems concerning cubic numbers are easy to state but are not so easy to solve. One such problem concerns the number of ways the number 3 can be expressed as the sum of three cubes. Mathematicians had known for centuries that the integer 3 can be expressed as the sum of three cubes in two ways, using relatively small numbers: $1^3 + 1^3 + 1^3 = 3$ and $4^3 + 4^3 + (-5)^3 = 3$. What mathematicians wanted to know is whether 3 can be expressed as the sum of three cubes in a third way.

Andrew Booker at Bristol University in the United Kingdom and Andrew Sutherland at the Massachusetts Institute of Technology in the United States tackled this problem and found the following solution in 2019:

$$(569,936,821,221,962,380,720)^3 + (-569,936,821,113,563,493,509)^3$$
$$+ (-472,715,493,453,327,032)^3 = 3$$

Booker and Sutherland found this solution by using a computer algorithm that was used on half a million idle computers that were made available for the task by half a million volunteers. The necessary amount of processing time to find the solution was equivalent to a single computer processor running continuously for 4 million hours, or more than 456 years.[8]

1   *Total sum of integers = 1 = 1³*

1

2 2

3   *Total sum of integers – 8 – 2³*

1

2 2

3 3 3

4 4   *Total sum of integers – 27 = 3³*

5

1

2 2

3 3 3

4 4 4 4   *Total sum of integers – 64 – 4³*

5 5 5

6 6

7

And so on.

The differences between successive cubes are 7, 19, 37, 61, 91, 127, and so on. These integers are part of the following interesting pattern:

$$2 + 3 + 2 = 7$$

$$3 + 4 + 5 + 4 + 3 = 19$$

$$4 + 5 + 6 + 7 + 6 + 5 + 4 = 37$$

$$5 + 6 + 7 + 8 + 9 + 8 + 7 + 6 + 5 = 61$$

And so on.

The prime numbers 7, 19, 37, 61, 91, 127, and so on are known as *cuban primes* because each of these primes is the difference between two successive cubic numbers. The German scientist Maximilian F. Hasler (1971–) gave a beautiful property of the cuban primes in the On-Line Encyclopedia of Integer Sequences (OEIS) on November 28, 2007. Hasler pointed out that the cuban primes form the following simple pattern:

$$7 \quad \rightarrow \quad 7 = 3\,(1 \times 2) + 1$$

$$19 \quad \rightarrow \quad 19 = 3\,(2 \times 3) + 1$$

$$37 \quad \rightarrow \quad 37 = 3\,(3 \times 4) + 1$$

$$61 \quad \rightarrow \quad 61 = 3\,(4 \times 5) + 1$$

And so on.

In the OEIS on September 15, 2005, Michael Somos mentioned that these cuban primes also satisfy the equation $4p = 1 + 3n^2$, where p is a cuban prime and n is a positive integer. Because of this equation, the following beautiful pattern emerges:

$$4 \times 7 = 28 = 1 + 3 \times 3^2$$

$$4 \times 19 = 76 = 1 + 3 \times 5^2$$

$$4 \times 37 = 148 = 1 + 3 \times 7^2$$

$$4 \times 61 = 244 = 1 + 3 \times 9^2$$

And so on.

My feeble attempts at trying to discover a pattern reveal that cuban primes also follow the following simple rule, where T_n denotes the nth triangular number:

$$7 + 2T_1 = 3^2$$

$$19 + 2T_2 = 5^2$$

$$37 + 2T_3 = 7^2$$

$$61 + 2T_4 = 9^2$$

And so on.

Can the sum of three *positive* cubes equal a cubic number? The answer is yes. The Indian mathematical genius Srinivasa Ramanujan (1887–1920) gave a formula for finding such cubes. There are an infinite number of solutions. The smallest solution is $3^3 + 4^3 + 5^3 = 6^3$. The second smallest is $7^3 + 14^3 + 17^3 = 20^3$. The third smallest is $37^3 + 30^3 + 27^3 = 46^3$.

Sums of consecutives cubes, beginning at a number other than 1, can sum to a square number, such as $25^3 + 26^3 + 27^3 + 28^3 + 29^3 = 315^2$ and $96^3 + 97^3 + 98^3 + 99^3 + 100^3 = 2,170^2$ and also $14^3 + 15^3 + 16^3 + 17^3 + 18^3 + 19^3 + 20^3 + 21^3 + 22^3 + 23^3 + 24^3 + 25^3 = 312^2$.

The sum of two cubed positive integers cannot equal a cubed integer. But close misses are possible. For example, $6^3 + 8^3 = 9^3 - 1$. Ramanujan succeeded in finding a formula that yields solutions to the equation $a^3 + b^3 = c^3 - 1$, where a, b, and c are positive integers.[9] Of course, if one transposes the -1 to the left-hand side of the equation and raises it to the power of 3, one obtains $a^3 + b^3 + 1^3 = c^3$. In other words, we obtain the sum of three positive cubic integers that equals a cubic integer. Two of the examples Ramanujan gave are $135^3 + 138^3 + 1^3 = 172^3$ and $11,161^3 + 11,468^3 + (-1)^3 = 14,258^3$.

Some deep results about cubes have been discovered over the years. For example, it has been shown that *almost* all cubes can be expressed as the sum of three cubes. Ryley's theorem states that every rational cube can be expressed as the sum of the cubes of three rational numbers.[10] (A rational number is a number that can be expressed as a fraction.) It is also known that any rational number, N, is the product of two sums of rational cubes. In other words, $(a^3 + b^3)(c^3 + d^3) = N$, where a, b, c, and d are rational numbers.

Integers that are the sum of two cubes are known as *taxicab numbers*. The reason why such integers acquired this name is interesting. The story is told that the outstanding British mathematician Godfrey Harold Hardy (1877–1947) traveled in 1918 to see his extremely ill friend and work colleague, Srinivasa Ramanujan, who was in a London hospital. Having arrived at Ramanujan's bedside, Hardy stated that he had traveled to the hospital in a taxicab that was numbered 1729. Hardy added that it appeared to him that 1729 is a very uninteresting number. However, Ramanujan strongly disagreed by saying, "No, Hardy! No, Hardy! The number 1729 is a very interesting number. It is the smallest number that can be expressed as the sum of two cubes in two ways: $1,729 = 1^3 + 12^3 = 9^3 + 10^3$." The story, which is believed to be true, dramatically reveals Ramanujan's intimacy with the integers. Hardy's work colleague

J. E. Littlewood (1885–1977), who was one of the best number theorists in the world at the time, referring to the Indian mathematician's marvelous insight into the properties of numbers, said that "all of the integers were Ramanujan's personal friends."

Three other examples of numbers that can be expressed as the sum of two cubes in two ways are

$$4{,}104 = 2^3 + 16^3 \text{ and } 9^3 + 15^3$$

$$13{,}832 = 2^3 + 24^3 \text{ and } 18^3 + 20^3$$

$$20{,}683 = 10^3 + 27^3 \text{ and } 19^3 + 24^3$$

The French mathematician Pierre de Fermat (1607–1665) proved that numbers that can be expressed as the sum of two positive cubes in n different ways exist for any n.

The smallest number that can be expressed as the sum of two cubes in three ways is

$$87{,}539{,}319 = 167^3 + 436^3 = 228^3 + 423^3 = 255^3 + 414^3$$

There are an infinite number of cubes that are palindromic. For example, the cube of 7 is 343, the cube of 11 is 1,331, and the cube of 101 is 1,030,301, all palindromes. The cube roots of these cubes are themselves palindromic. Is there any palindromic cube whose cube root is not palindromic? The answer is yes. It is an astonishing fact that of all the integers up to 2.8×10^{14}, only one palindromic cube has a cube root that is not palindromic. The palindromic cube 10,662,526,601 has a cube root equal to 2,201. Are there other palindromic cubes with non-palindromic cube roots? No one knows. As we go up the integers toward infinity, the probability of finding such a cube approaches zero. Charles R. Greathouse reported in the OEIS on May 16, 2011, that there are no other cubes with non-palindromic roots up to 10^{15}.

The infinite series of numbers that are the sum of two positive cubes is 2, 9, 16, 28, 35, 54, 65, 72, 91, 126, ... This series is remarkable because the first term, 2, is prime, but every other term is composite.

There are many interesting number curiosities involving cubic numbers. For example,

$$22 + 2 = 2^3 + 2^3 + 2^3$$

$$12 \times 3 = 1^3 + 2^3 + 3^3$$

$$32 \times 5 = 3^3 + 2^3 + 5^3$$

$$21{,}952 = (6 + 8 + 5 + 9)^3 \text{ and } 6{,}859 = (2 + 1 + 5 + 9 + 2)^3$$

$$244 = 1^3 + 3^3 + 6^3 \text{ and } 136 = 2^3 + 4^3 + 4^3$$

$$157 = 5^3 + 2^3 + 2^3 + 2^3 + 2^3 = 4^3 + 4^3 + 3^3 + 1^3 + 1^3$$

The digits of the cubic number 41,063,625 (345^3) can be rearranged to produce two other cubes: 56,623,104 (384^3) and 66,430,125 (405^3).

The numbers 1,375, 1,376, and 1,377 form the smallest set of three consecutive numbers that are divisible by cubes other than 1: 1,375/125 = 11, 1,376/8 = 172, and 1,377/27 = 51.

The two cubic numbers 8 and 125 are unique. They are the only two cubic integers that are four more than a square number: $8 = 2^2 + 4$ and $125 = 11^2 + 4$.

The number 27 is the only cube that is two more than a square number. Consequently, it can be shown that the set of the three consecutive numbers 12, 13, and 14 is unique in this sense: the sum of the first two terms, 12 and 13, equals a square. The sum of the second two terms, 13 and 14, equals a cube.[11]

It has been conjectured (but not yet proved) that the four cubic numbers 27, 64, 343, and 1,331 are the only cubes with two distinct digits that are not divisible by 10.

The positive integers 1, 8, 17, 18, 26, and 27 are the only positive numbers that are equal to the sum of the digits of their cubes. These cubes are known as *Dudeney numbers* after the English self-taught mathematician and puzzle genius Henry Ernest Dudeney (1857–1930) wrote about them.[12]

There are two possible sets of three consecutive integers whose cubes sum to a perfect square.[13] These two sets are $1^3 + 2^3 + 3^3 = 6^2$ and $23^3 + 24^3 + 25^3 = 204^2$.

The concatenation of two cubes cannot equal a cubic number.[14]

The numbers 153, 370, 371, and 407 are the only four positive integers that are equal to the sum of the cubes of their digits:

$$153 = 1^3 + 5^3 + 3^3$$

$$370 = 3^3 + 7^3 + 0^3$$

$$371 = 3^3 + 7^3 + 1^1$$

$$407 = 4^3 + 0^3 + 7^3$$

The only three consecutive *integers* whose sum equals a cube are:[15]

$$3^3 + 4^3 + 5^3 = 6^3$$

No complete discussion about cubes should omit the following discovery by Dudeney. He found two rational numbers other than 1 and 2 whose cubes sum to 9:

$$\left(\frac{415,280,564,497}{348,671,682,660}\right)^3 + \left(\frac{676,702,467,503}{348,671,682,660}\right)^3 = 9$$

The French mathematician Legendre had "proved" that 6 could not equal the sum of two cubes. However, Dudeney found the following simple solution:

$$\left(\frac{17}{21}\right)^3 + \left(\frac{37}{21}\right)^3 = 6$$

Many simple but interesting theorems involving cubic numbers have been found in number theory. For example, it can easily be shown that $n^3 - n$ (where n is an integer equal to or greater than 2) is always divisible by 6.

This initially surprising result yields readily to analysis. The factors of $n^3 - n$ are $(n - 1)$, n, and $(n + 1)$. These three factors are consecutives integers. One of the numbers in any three consecutive integers must be divisible by 2 and one of them by 3. Therefore, the product of three consecutive integers must be a multiple of 6. Subtracting 6 from this multiple of 6 leaves a number that is a multiple of 6. Consequently, $n^3 - n$ is always divisible by 6.

The formula for the partial sum of the cubes, $1^3 + 2^3 + 3^3 + 4^3 + 5^3 + \ldots + n^3$, is

$$1^3 + 2^3 + 3^3 + 4^3 + \ldots + n^3 = \left(\frac{n(n + 1)}{2}\right)^2$$

The formula for the partial sum of the *odd* cubes, $1^3 + 3^3 + 5^3 + \ldots + (2n - 1)^3$, is

$$1^3 + 3^3 + 5^3 + \ldots + (2n - 1)^3 = n^2(2n^2 - 1)$$

The odd cubic numbers are closely related to the square root of 2 in a surprising way. The convergents of the square root of 2 are 7/5, 17/12, 41/29, 99/70, 239/169, ... Beginning with the first convergent, 7/5, the square of the product of the two numbers in every *second* convergent equals the partial sum of the *odd* cubes, beginning with 1^3.

Thus,

$$1^3 + 3^3 + 5^3 + 7^3 + 9^3 = (7 \times 5)^2$$

$$1^3 + 3^3 + 5^3 + 7^3 + 9^3 + \ldots + 57^3 = (41 \times 29)^2$$

$$1^3 + 3^3 + 5^3 + 7^3 + 9^3 + \ldots + 168^3 = (239 \times 169)^2$$

And so on.

Finally, here are two problems concerning cubic numbers:

1. Can the reader prove that the only prime number that is one less than a cube is 7?
2. Can the reader prove that the only cubic number that is also triangular is 1?

Solutions

1. We were asked to prove that 7 is the only prime that is one less than a cube. Let p equal a prime number that is one less than n^3, a cubic number. Thus, $p = n^3 - 1$. If we factor the right-hand side of the equation, we obtain

$$p = (n - 1)(n^2 + n + 1)$$

Because p is a prime, the only factors of p are 1 and p itself. Therefore, $n - 1$ must equal 1, and $n^2 + n + 1$ equals p. Since $n - 1 = 1$,

$$n = 2$$

Therefore, $n^2 + n + 1$ must equal 7. Thus, $p = 7 = 2^3 - 1$.

2. We were asked to prove that the only cubic number that is also triangular is 1. This is easily proved. A triangular number, T_n, is given by the formula $\frac{1}{2} n (n + 1)$. If this equals a cubic number, we can write

$$\tfrac{1}{2} n (n + 1) = m^3$$

Multiply both sides of the equation by 8 to obtain

$$4n^2 + 4n = 8m^3$$

Since $8m^3 = (2m)^3$, we can write

$$4n^2 + 4n = (2m)^3$$

or

$$(2n + 1)^2 - 1 = (2m)^3$$

or

$$(2n + 1)^2 - (2m)^3 = 1$$

According to Catalan's theorem, the only two powers that differ by 1 are 3^2 and 2^3. Therefore, $2n + 1$ must equal 3, which means that n is equal to 1. Also, $(2m)^3$ equals 8; therefore, m also equals 1. Consequently, $(2n + 1)^2 - (8m)^3$ equals $9 - 8 = 1$. Thus, the original equation $1/2\, n (n + 1) = m^3$ must equal $1/2\, 1(1 \times 2) = 1^3$. Therefore, the only cubic number that is also triangular is 1.

CHAPTER 3

Some Words Concerning the Square Root of 2

When I was a schoolboy, I was astonished when I first learned that the square root of 2 is an irrational number. Some students in my class may have taken this fact in their stride, but I was in awe when I learned that the square root of 2 could not be expressed as a fraction.

Many years have passed since my school days, but I am still astonished today by the irrational numbers. These are numbers that cannot be expressed as a ratio between two integers. At the time, I thought it was inconvenient that such numbers should exist. After all, it would have been much simpler if all numbers could be expressed as a ratio between two integers.

In those far-off days, I found it extremely strange that such numbers exist. I still do. This early discovery, that irrational numbers exist, demonstrated to me that the results of mathematics were not imposed by human beings on the mathematical structure. On the contrary. It seemed to me that we human beings had to accept mathematical results even if those results were inconvenient to us. We might not like irrational numbers, but we had to accept that they exist because mathematical reality is built that way. This was a form of proof to me at that early age that we human beings *discover* mathematics. We do not invent it.

I am not sure if schoolchildren (or indeed adults) today realize the weird nature of the irrationals.

Here is one way of getting a student of mathematics to appreciate the wonder of irrational numbers. Explain to the student that between any two fractions on the number line, there are an infinite number of points that can be expressed as a fraction. For example, consider the gap between 3/4 and 1/2. Is there a number between those two points on the number line that can be expressed as a fraction? Yes, there is: 5/8. Is there a number that can be expressed as a fraction between 5/8 and 1/2? Yes, there is: 9/16. By proceeding along these lines, you will soon convince the student that there is an infinite number of points on the number line that can be expressed as a fraction.

But then you spring a pleasant and amazing surprise: in addition to the infinite number of numbers on the number line that can be expressed as a fraction, there are also an infinite number of numbers on the number line that *cannot* be expressed as a fraction. These latter numbers are known as irrational numbers. Because these numbers are irrational, their decimal expansion goes on forever.

One of the most famous irrational numbers (and the first to be discovered) is the positive square root of 2, usually written as $\sqrt{2}$, or $2^{1/2}$. By its very definition, this is a number that, when multiplied by itself, *exactly* equals 2. It is often referred to as the *principal square root* of 2 to distinguish it from the negative square root of 2. The positive square root of 2 occurs naturally as the length of the diagonal in a square whose side has a length of 1. Because the $\sqrt{2}$ is irrational, the length of the diagonal in a unit square can never be measured *exactly*. The length of the diagonal is said to be *incommensurable* to the length of the side of the square.

The value of the square root of 2 to the 20th decimal place is 1.41421356237309504880.

The fraction 17/12 is close to the value of the $\sqrt{2}$. If that fraction (or other similar fraction) exactly equaled the $\sqrt{2}$, we could, of course, measure exactly the length of the diagonal of any square. But no such fraction exists. Consequently, it is an astonishing fact that although the diagonal of any square may be staring one in the face—and one assumes that the length of the side of that square is an exact number—the length of the diagonal of such a square can never be measured exactly. Not now. Not ever.

There have been many proofs published that the $\sqrt{2}$ is an irrational number. This means that the decimal expansion of the $\sqrt{2}$ goes on forever. Some of these proofs have been based on geometry. Others have been based on simple number theory.

All rational numbers can be expressed as a finite continued fraction. Irrational numbers can be expressed as an *infinite* continued fraction.

The irrationality of the square root of 2 is believed to have been discovered by the Pythagoreans about 2500 years ago. The limit of the following infinite continued fraction is the $\sqrt{2}$:

$$\sqrt{2} = 1 + \cfrac{1}{2 + \cfrac{1}{2 + \cfrac{1}{2 + \cfrac{1}{2 + \ldots}}}}$$

If this continued fraction is truncated at different points, it gives a fraction that closely *approximates* the $\sqrt{2}$. For example,

$$\sqrt{2} \approx 1 + \frac{1}{2} = \frac{3}{2} = 1.5$$

$$\sqrt{2} \approx 1 + \cfrac{1}{2 + \cfrac{1}{2}} = \frac{7}{5} = 1.4$$

$$\sqrt{2} \approx 1 + \cfrac{1}{2 + \cfrac{1}{2 + \cfrac{1}{2}}} = \frac{17}{12} = 1.416\ldots$$

$$\sqrt{2} \approx 1 + \cfrac{1}{2 + \cfrac{1}{2 + \cfrac{1}{2 + \cfrac{1}{2}}}} = \frac{41}{29} = 1.413\ldots$$

$$\sqrt{2} \approx 1 + \cfrac{1}{2 + \cfrac{1}{2 + \cfrac{1}{2 + \cfrac{1}{2 + \cfrac{1}{2}}}}} = \frac{99}{70} = 1.41428$$

The above *infinite* continued fraction is also a proof that the $\sqrt{2}$ is irrational. If the $\sqrt{2}$ were rational, its continued fraction would be *finite*.

The further down the continued fraction one truncates, the better approximations one obtains to the $\sqrt{2}$. The fractions obtained in this manner are 3/2, 7/5, 17/12, 41/29, 99/70, 239/169, 577/408, 1,393/985, and so on. This series of fractions is named *Eudoxus' ladder* in honor of an ancient Greek mathematician named Eudoxus (390–337 BCE), who was a student of Plato's. Because of the way the fractions converge on the square root of 2, the following identities hold in relation to every second successive denominator, beginning with the denominator of the second fraction, 7/5:

$$\frac{5}{2} - \frac{2}{5} = 2.1$$

$$\frac{29}{12} - \frac{12}{29} = 2.0028\ldots$$

$$\frac{169}{70} - \frac{70}{169} = 2.000084\ldots$$

This series of fractions approaches the value of 2 the further out the series we go.[1] This second series also converges on 2 but from the other side as we go out the series:

$$\frac{7}{3} - \frac{3}{7} = 1.9047\ldots$$

$$\frac{41}{17} - \frac{17}{29} = 1.9971$$

$$\frac{239}{99} - \frac{99}{239} = 1.9999\ldots$$

The series that is formed from the denominators of Eudoxus' ladder is 0, 1, 2, 5, 12, 29, 70, 169, 408, 985, ... These numbers are usually referred to as the Pell numbers. One property of the numbers in this series is that twice the square of each term in the series is one more or one less than a square number.

The numerators in Eudoxus' ladder, 3, 7, 17, 41, 99, . . . , have many interesting properties also. Here is one little-known property:

$$\left(1 - \sqrt{2}\right)^{2} = 3 - 2\sqrt{2}$$

$$\left(1 - \sqrt{2}\right)^{3} = 7 - 5\sqrt{2}$$

$$\left(1 - \sqrt{2}\right)^{4} = 17 - 12\sqrt{2}$$

$$\left(1 - \sqrt{2}\right)^{5} = 41 - 29\sqrt{2}$$

$$\left(1 - \sqrt{2}\right)^{6} = 99 - 70\sqrt{2}$$

And so on.

Here are two short and simple proofs that the $\sqrt{2}$ is irrational.

The first proof is based on the simple fact that every positive integer can be expressed as $3x \pm 0$, $3x \pm 1$, $3x \pm 2$. Hence, the last *nonzero* digit of any square number expressed in base 3 must be 1, whereas the last *nonzero* digit of twice a square expressed in base 3 must be 2. Therefore, an equation in integers satisfying $a^2 = 2b^2$ cannot exist. This proves that a fraction equaling the $\sqrt{2}$ cannot exist.[2]

The second proof we offer is a geometrical proof of the irrationality of the $\sqrt{2}$. If the $\sqrt{2}$ is rational, one can find the *smallest* possible isosceles triangle with integer sides. (See the figure on page 23.) Circumscribe a circle whose radius is equal to the vertical side of the triangle. Construct the tangent on the hypotenuse. This process has now created a *smaller* isosceles triangle with integer sides that are smaller than the smallest isosceles triangle that we assumed we had initially found. This is a contradiction. Therefore, our original assumption, that the $\sqrt{2}$ is rational, cannot be correct. The $\sqrt{2}$ must therefore be irrational:

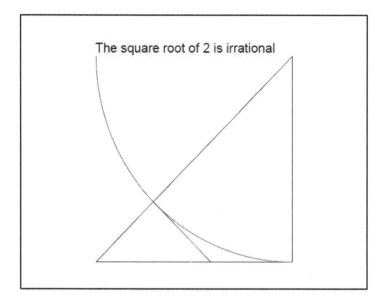

The square root of 2 is irrational

The $\sqrt{2}$ crops up in a famous proof that there exist two irrational numbers, say, a and b, such that a^b is rational. Let $a = b = \sqrt{2}$. There are two possible results; either the resulting number is rational, or it is irrational. Let's consider the first possible case. If the $\sqrt{2}$ raised to the power of the $\sqrt{2}$ is rational, we have found an irrational number raised to an irrational power that is rational. That solves the problem.

However, suppose the $\sqrt{2}$ raised to the power of the $\sqrt{2}$ is irrational. If that is the case, we then raise *that* number to the power of $\sqrt{2}$. Thus, we have $a = \sqrt{2}^{\sqrt{2} \times \sqrt{2}}$, which makes $a = \sqrt{2}^2$. Thus, $a = 2$. Therefore, in either case, we have found an irrational number raised to an irrational power that equals a rational number.

As it happens, the number $\sqrt{2}^{\sqrt{2}}$ is transcendental, according to a theorem in mathematics known as the Gelfond–Schneider theorem. It equals 2.665144142. . . . The number was originally proved transcendental in 1930 by the Russian mathematician Rodion Osievich Kuzmin (1891–1949). It was later independently proved transcendental by the Soviet mathematician Aleksandr Gelfond (1906–1968) and the German mathematician Theodor Schneider (1911–1988), who discovered the Gelfond–Schneider theorem, from which the result that Kuzmin obtained immediately follows.

The $\sqrt{2}$ makes an appearance in the square root of the imaginary number i and in the square root of $-i$. Students of recreational mathematics find this a pleasant surprise when they first learn of it. The so-called imaginary number i is defined so that $i^2 = -1$. Also, $-i^2 = -1$. Why, they ask, should an irrational number, such as the $\sqrt{2}$, be an integral part of both the square root of i and the square root of $-i$? Fortunately, it is not difficult to see why.

The square root of i is

$$\frac{(1 + i)}{\sqrt{2}} = \sqrt{i}$$

The square root of $-i$ is

$$\frac{-(1 + i)}{\sqrt{2}} = \sqrt{-i}$$

Readers are urged to square both quantities on the left-hand side of these two equations to satisfy themselves that they equal i and $-i$, respectively.

Readers will recall that imaginary numbers define complex numbers, which were first used to solve certain types of quadratic and cubic equations in the sixteenth century. Today, those studying electrical engineering, electromagnetism, physics, number theory, and geometry all use complex numbers. Most mathematicians today recognize that the so-called imaginary numbers are as real as all other numbers.

The only positive real number (other than 1) that has an infinite tower of itself created by repeated exponentiation and that is exactly equal to its square is the $\sqrt{2}$. In other words,

$$\left(\sqrt{2}\right)^{\sqrt{2}^{\sqrt{2}^{\sqrt{2}^{\sqrt{2}^{\cdots}}}}} = 2.$$

Mathematicians have discovered various ways in which the $\sqrt{2}$ can be expressed as the sum of an infinite series. These series are beautiful and illustrate the fundamental value of the $\sqrt{2}$. Here is one such series:

$$\sqrt{2} = 1 + \frac{1}{2} - \frac{1}{2 \times 4} + \frac{1 \times 3}{2 \times 4 \times 6} - \frac{1 \times 3 \times 5}{2 \times 4 \times 6 \times 8} + \cdots$$

The reciprocal of the $\sqrt{2}$ can also be expressed as the sum of an infinite series. For example, as k goes from 0 toward infinity, the following series converges on the reciprocal of $\sqrt{2}$:

$$\frac{1}{\sqrt{2}} = 1 - \frac{1}{(4k + 2)^2} = \left(1 - \frac{1}{4}\right)\left(1 - \frac{1}{36}\right)\left(1 - \frac{1}{100}\right)\left(1 - \frac{1}{196}\right)\cdots$$

The $\sqrt{2}$ has several interesting properties. For example, the reciprocal of the $\sqrt{2}$ is equal to one-half of the $\sqrt{2}$. Mathematically, we can write this identity as

$$\frac{1}{\sqrt{2}} = \frac{\sqrt{2}}{2}$$

Another interesting equality is

$$(\sqrt{2} + 1) = \frac{1}{(\sqrt{2} - 1)}$$

As one would expect, the $\sqrt{2}$ crops up in several places in geometry. However, it also appears in unexpected places. For example, if the surface area of a sphere is 8π units, the radius of the sphere equals the $\sqrt{2}$.

The $\sqrt{2}$ equals 1.41421356237309. . . . In *The Call of the Primes* (2016, p. 162), I give the following curiosity concerning the first nine decimals of the $\sqrt{2}$.

Write the first nine decimals in the form of an addition sum like this:

$$
\begin{array}{r}
414 \\
213 \\
\underline{562} \\
1{,}189
\end{array}
$$

Curiously, the numbers sum to 1189, which equals 41 × 29. Earlier in this chapter, we showed that the fourth fraction in Eudoxus' ladder is 41/29.

Here is another curiosity that I did not include in *The Call of the Primes*, where the digits of the number 1,189 crop up again. The square root of the square root of 2 equals 1.189. . . . In other words, $\sqrt[4]{2} = 1.189\ldots$.

Is there a simple way to generate the fractions in Eudoxus' ladder? Yes, there are several.

Here is one method.

Write the following series of integers (beginning with 1), where twice each number in the sequence plus the previous number equals the next number. Below that first series, write the differences between consecutive integers in the first series:

$$
\begin{array}{ccccccccc}
1 & 2 & 5 & 12 & 29 & 70 & 169 & 408 \ldots \\
& 1 & 3 & 7 & 17 & 41 & 99 & 209 &
\end{array}
$$

Pair each number in the second series (beginning with 3) with the number above it and immediately to the left in the first series to form a series of fractions as follows: 3/2, 7/5, 17/12, 41/29, 99/70, and so on. This is the series of successive convergents given earlier in this chapter as the rungs of Eudoxus' ladder.

Finally, here is a little problem that the reader may enjoy solving. Consider the following equation:

$$
\sqrt[\sqrt{2}]{\sqrt{2}} = \sqrt{2}^{\sqrt{2}}
$$

The problem is, is this equation true? The solution is given at the end of this chapter.

Addendum

The following curiosity concerning the square root of 2 was discovered by the German mathematician Professor Ronald Percival Sprague (1894–1967).[3]

Write in a line the successive multiples of the square root of 2, ignoring the fractional parts. Beneath this sequence, write the integers missing from the first series. One obtains

1	2	4	5	7	8	9	11	12 ...

3	6	10	13	17	20	23	27	30

The difference between the top and bottom integers at the nth position is always equal to $2n$.

It appears to have gone unnoticed in the mathematical community that a similar and perhaps more fundamental property applies to the multiples of phi:

phi equals 1.6180339. . . .

Write in a line the multiples of phi, ignoring the fractional parts. Beneath this list of integers, write down the integers that do not appear in the first list. One obtains

1	3	4	6	8	9	11	12	14	16	17	19	21	22	24

2	5	7	10	13	15	18	20	23	26	28	31	34	36	39

The difference between the bottom integer and the top integer at the nth place is n.

Solution

We were asked if the following equation is true.

$$\sqrt[\sqrt{2}]{\sqrt{2}} = \sqrt{2}^{\sqrt{2}}$$

Yes, the equation is true.

To see this, consider the left-hand side of the equation. We find that the 1.4142135th root of the $\sqrt{2}$ equals 1.6325269, correct to seven significant decimals. You will find that 1.6325269 . . . raised to the power of 1.4142135 . . . equals 2.

The right-hand side of the equation is easily calculated with the help of a scientific calculator. It equals 1.6325269, correct to seven significant decimals. Thus, both sides of the equation equal 1.6325269, correct to seven significant decimal places.

CHAPTER 4

Waring's Problem and Other Related Matters

Mathematical enthusiasts sometimes do a little doodling with pen and paper and play with the integers. In doing so, one might notice that, say, 13 can be expressed as the sum of two squares $(2^2 + 3^2)$; the number 14 can be expressed as the sum of three squares $(2^2 + 3^2 + 1^2)$, and 15 can be expressed as the sum of four squares $(2^2 + 3^2 + 1^2 + 1^2)$.

The math enthusiast may then ask herself if there is any integer that cannot be expressed as the sum of at most four squares? She will find that there is not. She will, however, find that most integers can be expressed as the sum of three squares. Only integers that are equal to $8n + 7$, where n is an integer, require four squares summed to equal them. Thus, 7 requires four squares, 15 requires four squares, 23 requires four squares, 31 requires four squares, and so on.

The math enthusiast may then consider expressing numbers as the sum of cubes. She will find that 9 requires two cubes $(2^3 + 1^3)$ to be summed to equal it. She will find that 12 equals five cubes $(2^3 + 1^3 + 1^3 + 1^3 + 1^3)$ to be expressed. Continuing in this manner, she will find that 23 requires nine cubes $(2^3 + 2^3 + 1^3 + 1^3 + 1^3 + 1^3 + 1^3 + 1^3 + 1^3)$. She may wonder if any integer requires more than nine cubes summed to express it. The answer is no. It is a remarkable fact that of the infinite set of positive integers, only two integers, 23 and 239, require nine cubes to be summed to express them.

The mathematical enthusiast may then consider how many fourth powers, fifth powers, sixth powers, and so on are needed to sum to any number, k.

Professional mathematicians approach mathematics in a comparable way to our imaginary math enthusiast. They obtain certain results in one specific area. They then try to prove that specific result. If they succeed, they will then try to generalize the problem and try to obtain a general formula to the problem at hand.

The problem as stated with our imaginary math enthusiast leads one to what is known as *Waring's problem*. The problem, which appeals to recreational mathematicians, is usually stated as follows: can every positive integer N be represented as a sum of s kth powers of positive integers, where s does not depend on N? It is a famous problem in mathematics. The general solution to Waring's problem is still unknown.

Waring's problem has engrossed mathematicians for centuries. The modern formulation of the problem was first proposed by the English mathematician Edward Waring (1736–1798) in 1770. Waring proposed that every integer is the sum of a fixed number $g(k)$ (called *little gee*) of kth positive powers, which depends only on the value of k. Waring conjectured that $g(2)$ is 4, $g(3)$ is 9, and $g(4)$ is 19. In other words, Waring conjectured that every integer can be expressed as the sum of at most four squares, as the sum of at most nine cubes, and as the sum of at most 19 fourth powers. During the twentieth century, these conjectures were proved to be true.

In general, it is not entirely obvious that every integer is the sum of a fixed number $g(k)$ of kth powers, which depends only on the value of k. It was as recent as 1909 that the German mathematician David Hilbert (1862–1943) proved that $g(k)$ did in fact exist for every integer.

Waring's conjecture was not entirely new. In 1770, the Italian-born French mathematician Joseph-Louis Lagrange (1736–1813) had proved that every integer is the sum of at most four square numbers. Lagrange's proof was anticipated in the third century by Diophantus of Alexandria, who conjectured that every integer could be expressed as the sum of at most four square integers.

In 1912, the German mathematician Arthur Wieferich (1884–1954) and the English mathematician Aubrey Kempner (1880–1973) proved that $g(3)$ is 9. Seventy-four years later, in 1986, the Indian mathematician Ramachandran Balasubramanian (1951–) and the two French mathematicians Jean-Marc Deshouillers (1946–) and François Dress (?), all working together, proved that $g(4)$ equals 19. Before that, in 1964, the Chinese mathematician Chen Jingrun (1933–1996) proved that $g(5)$ is 37. (It is relatively unknown that the English mathematician John Horton Conway [1937–2020] independently proved the case for $g(5)$ while he was still a student.) In 1946, the Indian mathematician S. S. Pillai (1901–1950) proved that $g(6)$ is 73.

Around 1772, J. A. Euler, the son of the famous mathematician Leonhard Euler (1707–1783), *conjectured* that $g(k)$ is at least as large as $2^k + [(3/2)^k] - 2$. (The symbols [ ] represent the floor function, which tells us to take the greatest integer less than the value between the floor function symbols. Thus, [4.5] equals 4 and so on.) When k equals 1, the formula gives 1. That is correct. Every number requires the sum of one power, namely, itself, raised to the power of 1, to express itself as the sum of powers. When n equals 2, the formula gives 4. In other words, the sum of at least four squares is required to express all numbers. When n equals 3, the formula gives 9. In other words, the sum of at least nine cubes is required to express all numbers. The formula known to J. A. Euler is known as a lower bound formula because it gives the least number of powers required to equal $g(k)$.

The formula that was known to J. A. Euler became widely known to other mathematicians. Many of those mathematicians *proved* that $g(k)$ is at least as large as $2^k + [(3/2)^k] - 2$. Over time, many mathematicians began to suspect that the formula is an *exact* formula for $g(k)$. In recent decades, they noticed that the formula gave the correct results for $g(2)$, $g(3)$ and $g(4)$, and $g(5)$ and $g(6)$. In other words, they suspect that $g(k) = 2^k + [(3/2)^k] - 2$ for *all* positive integer values of k. To this day, however, this suspicion has never been proved. It is the heart of the puzzle which mathematicians call Waring's problem.

The first 10 values given by the formula for $g(k)$ are 1, 4, 9, 19, 37, 73, 143, 279, 548, and 1,079.[1]

There is a second part to what is known as Waring's problem that concerns itself with $G(k)$.

The quantity called $G(k)$ (called *big gee*) has also been studied by mathematicians. $G(k)$ is defined as the least possible integer, s, such that every sufficiently large integer (i.e., every integer greater than some constant) can be represented as a sum of at most s positive integers to the power of k. Obviously, $G(1) = 1$. We know that integers of the form $8n + 1$ cannot be expressed as the sum of three squares. Therefore, $G(2) = 4$. Mathematicians have over the years proved that $G(4)$ is 12. In other words, there is some integer, x, such that every integer beyond x can be expressed as the sum of 12 fourth powers. This is the only *known* exact value for $G(k)$.

We know that every number requires the sum of at most nine cubes to represent it. But mathematicians have now discovered that some numbers may require as little as the sum of four cubes to represent them. Thus, it *may* be the case that $G(3) = 4$, but this has not yet been proven. It is conjectured that the largest number known that is not the sum of four cubes is 7,373,170,279,850.[2]

There are only 15 numbers that require the sum of eight cubes to represent them. Curiously, one of those numbers is 15. The full list of those 15 numbers is 15, 22, 50, 114, 167, 175, 186, 212, 231, 238, 303, 364, 420, 428, and 454.[3] This means, of course, that $G(3) \leq 7$. The largest number requiring the sum of seven cubes to represent it is 8,042.[4]

The number 13,792 requires the sum of 17 fourth powers to represent it. It was shown in the year 2000 that every number between 13,793 and 10^{245} can be expressed as the sum of 16 fourth powers. Later, it was proved that every integer beyond 10^{220} can be expressed as the sum of 16 fourth powers. Is there a least possible integer, x, such that every integer larger than x can be represented as the sum of at most 15 fourth powers? The answer is no. All numbers of the form 31×16^n require the sum of 16 fourth powers to represent them. Thus, $G(4) = 16$.

It is conjectured that every integer (positive or negative) that is not congruent to ± 4 modulo 9 can be expressed as the sum of three (positive or negative) cubes. (The cubes, modulo 9, are 0, 1, and -1. Therefore, the sum of any three such numbers cannot be congruent to ± 4 modulo 9.)

Computer searches in recent years have found all the integers less than 100 (that are not equal to ± 4 modulo 9) that can be expressed as the sum of three positive or negative cubes. The smallest integer that cannot be expressed as the sum of three positive or negative cubes is 114.

Recreational math enthusiasts often play with the powers of integers for fun. In doing so, they often find many pleasant surprises. For example, consider the above problem of finding the smallest solutions of expressing integers as the sum of three positive or negative cubes. One will find that the smallest solution for $15 = 2^3 + 2^3 - 1^3$ and the smallest solution for $17 = 2^3 + 2^3 + 1^3$. What about 16? Can 16 be expressed as such a sum? Yes. But the smallest solution is $1{,}626^3 + (-1{,}609)^3 + (-511)^3$.

One will find that the smallest such solutions for 27 are $6^3 + (-5^3) + (-4^3)$, $28 = 3^3 + 1^3 + 0^3$, and $29 = 3^3 + 1^3 + 1^3$. One would be inclined

to think that 30 can also be expressed as the sum of three positive or negative cubes in similar relatively small integers. But the smallest solution is
$30 = 2{,}220{,}422{,}932^3 + (-283{,}059{,}965)^3 + (-2{,}218{,}888{,}517)^3$.

Consider the representation of 3 as the sum of three positive cubes. Most schoolchildren aged twelve or slightly more know that $3 = 1^3 + 1^3 + 1^3$. Slightly older children may know the following result: $3 = 4^3 + 4^3 + (-5)^3$. For more than 60 years, these were the only two ways that mathematicians knew that express 3 as the sum of three cubes. In 2019, Andrew Booker and Andrew Sutherland found a third representation of 3:

$$3 = (569{,}936{,}821{,}221{,}962{,}380{,}720)^3 + (-569{,}936{,}821{,}113{,}563{,}493{,}509)^3$$
$$+ (-472{,}715{,}493{,}453{,}327{,}032)^3.$$

We all know that $2 = 1^3 + 1^3 + 0^3$.

But how many know that $2 = (1{,}214{,}928)^3 + (3{,}480{,}205)^3 + (-3{,}528{,}875)^3$?

Students of recreational mathematics are usually interested in the sum of the powers of consecutive integers, beginning with 1. Such students may find the five following formulas, taken from the vast literature on this topic, interesting.

For any positive integer n,

$$1 + 2 + 3 + \ldots + n = \frac{n(n + 1)}{2}$$

$$1^2 + 2^2 + 3^2 + \ldots + n^2 = \frac{n(n + 1)(2n + 1)}{6}$$

$$1^3 + 2^3 + 3^3 + \ldots + n^3 = \frac{n^2(n + 1)^2}{4}$$

$$1^4 + 2^4 + 3^4 + \ldots + n^4 = \frac{6n^5 + 15n^4 + 10n^3 - n}{30} = \frac{n^5}{5} + \frac{n^4}{2} + \frac{n^3}{3} - \frac{n}{30}$$

$$1^5 + 2^5 + 3^5 + \ldots + n^5 = \frac{2n^6 + 6n^5 + 5n^4 - n^2}{12}$$

CHAPTER 5

Drawing Conclusions from Mathematical Data Can Sometimes Lead to Error

Suppose someone wrote the following series on a piece of paper and showed it to you: 1, 2, 3, 4, 5, 6, 7, 8, 9. . . . They then ask you which number comes after 9 in the series. You will probably answer 10. Although 10 is a correct answer, it is not the *only* correct answer. The series of numbers could look like this, for instance: 1, 2, 3, 4, 5, 6, 7, 8, 9, 23, 49, 61, 100, 122, 136, 148, 163. . . .

This answer is also correct if the series consists of numbers such that raising the integers of the digits in the number to the power of 2 results in a square number being formed. This series of numbers may be found on the On-Line Encyclopedia of Integer Sequences, where it is designated series number A048386.

This simple example illustrates that one needs to be careful before drawing conclusions from mathematical data.

Many famous mathematicians have incorrectly formed conclusions from a set of mathematical data. For example, consider the following conjecture made by the Hungarian mathematician George Pólya (1887–1985). It states that for any given positive integer n, where n is greater than 1, at least half of the positive integers less than or equal to n have an odd number of prime factors. The conjecture was first stated by Pólya in 1919.

The conjecture was proved false in 1958.

George Pólya is believed to have checked the first 1,500 positive integers. No counterexample to the conjecture was found. Thus, all these instances (in the mind of Pólya) increased the likelihood that the conjecture was true.

Over the years, mathematicians inspected various integers, but no exceptions to Pólya's conjecture were found. Some felt that these absences of exceptions were strong grounds for believing that the conjecture was true.

However, the conjecture was shown to be false by Brian Haselgrove in 1958. Haselgrove believed that the first counterexample of the conjecture occurred around 101.845×10^{361}. In 1960, R. Sherman Lehman, professor emeritus at the University of California, Berkeley, discovered an exact integer, 906,180,359, which contradicted the conjecture. In 1980, Minoru Tanaka of Gakushuin University, Tokyo, discovered the smallest counterexample, 906,150,257.[1]

Christian Goldbach (1690–1764) made a famous conjecture that every even number greater than two is the sum of two primes. That conjecture is still unproved, although it is reasonable to say that most mathematicians today believe the conjecture is true. Goldbach also made the following conjecture: every odd composite number is the sum of a prime and twice a square number. Thus, $9 = 7 + 2 \times 1^2, 15 = 7 + 2 \times 2^2, 21 = 3 + 2 \times 3^2$, and so on. As one goes up the list of odd composite integers, one finds that the odd composites, one after another, can apparently be expressed as the sum of a prime plus twice a square. It becomes relatively easy to accept that perhaps this is always the case.

However, the second Goldbach conjecture is false.

The odd composite 5,777 (its prime factors are 53 and 109) cannot be expressed as the sum of a prime number plus twice a square. The odd composite 5,993 (its prime factors are 13 and 461) also cannot be expressed as the sum of a prime plus twice a square. These are the only known two odd composites that cannot be expressed as the sum of a prime plus twice a square. A computer search up to 4,000,000,000 failed to find a counterexample.[2]

Many readers may think that it was easy for mathematicians in the nontechnological distant past to make conjectures and believe that today, with the possibility of computer searches, probably no one would make any such conjecture. Alas, that is unlikely to be true. The story of number theory is filled with conjectures. Some have been proved to be true. Some are still unproven. Some, of course, have been proved to be false.

For example, consider the following problem, which arises if we partition the primes into two groups: primes of the form $4n + 3$ and primes of the form $4n + 1$. One might initially suspect that the primes would be split rather evenly between the two groups. But as one checks the primes up to some integer x, this does not appear to be the case. If one inspects the distribution of the primes, beginning at 3, one will find that there appears to be more primes of the form $4n + 3$ than there are primes of the form $4n + 1$. As one goes up the numbers, say, from 1,000 to 5,000 to 10,000 to 20,000, one still finds that there are (slightly) more primes of the form $4n + 3$ than there are of $4n + 1$.

In 1853, Chebyshev noticed this imbalance between the numbers of each type of prime, and since then, the phenomenon has become known as Chebyshev's bias. Chebyshev conjectured that, up to any given number x, there are always more primes of the form $4n + 3$ than of the form $4n + 1$.

However, the conjecture is false.

The conjecture first fails at the prime number 26,861. At that point, there are a greater (by one) number of primes of the form $4n + 1$. But since 26,863 is also prime, at that point, the number of the two forms of primes is even again. It turns out that the lead in the race between the two forms of primes swings back and forth infinitely often as one goes up the numbers toward infinity. The great English number theorist John Edensor Littlewood (1885–1977) proved this in a theorem in 1914.

Curiously, most of the time, up to any given number x, there are more primes of the form $4n + 3$ than there are of the form $4n + 1$. Yes, there are times when the lead is taken by primes of the form $4n + 1$. But from inspection, such leads are usually short lived. Suppose one chooses a large integer x and then checks how many primes of the form $4n + 3$ exceeds those of $4n + 1$ up to x. Repeat this process many times, increasing the value of x at each stage. Empirically, there are always more primes of the form $4n + 3$ than of the form $4n + 1$. But the race changes infinitely often. Which form of prime occurs most often as x goes toward infinity? There are different types of measurements used by mathematicians to determine this answer. According to one such measurement, the primes of the form $4n + 3$ occur 99.59 percent of the time as the value of x approaches infinity.[3]

As one inspects the primes, one finds that there are other surprises. To see one of these, let us partition the primes into the following two groups: primes of the form $3n + 2$ and primes of the form $3n + 1$. The first group of primes consists of 2, 5, 11, 23, 29, 41, ... The second group of primes consists of 7, 13, 19, 31, 37, 43, ... Are there always more primes of the $3n + 2$ variety up to any given number x? The answer is no. Littlewood's theorem applies here also. The race between the two forms of primes changes infinitely often.

But suppose one did not know of Littlewood's theorem. If that individual today programmed a computer to search for both kinds of primes, one would initially find, as one goes up the number line, that there are more primes of the form $3n + 2$ than there are primes of the form $3n + 1$. As the computer program searches first in the tens of thousands of integers, then in the tens of millions, and then in the tens of trillions, one will see that there are more primes of the form $3n + 2$. But eventually, the tide turns in accordance with Littlewood's theorem, and there are more primes of the form $3n + 1$ than of those of the form of $3n + 2$. The first time this happens is at the prime number 608,981,813,029.[4]

Finally, we briefly mention another famous prime race. In mathematics, there is a function known as the *logarithmic integral function*, usually symbolized as $\text{li}(x) = R \times 0\, dt / \log t$. This function is used to approximately count how many primes are less than any given integer x. There is also another function, also used to approximately count the primes less than any given integer x. This second function is symbolized as function $\pi(x)$. As we go up the primes, the function $\text{li}(x) = R \times 0\, dt / \log t$ always seems to be larger than $\pi(x)$. The question naturally arises: does the second function ever give an approximation that is larger than that given by the first function? Littlewood proved that it does and that this change in the race between the two functions occurs infinitely often. But we do not know at what number this change in the race first occurs. It is believed to be in the range 1.4×10^{316}.[5]

Summing up, it is astonishing, when one thinks about it, that the simple prime numbers, 2, 3, 5, 7, 11, 13, 17 . . . , could generate such problems that the greatest minds on Earth find difficult to grapple with. But, as we have seen, such is the case. The great Swiss mathematician Leonhard Euler said, "Mathematicians have tried in vain to this day to discover some order in the sequence of prime numbers, and we have reason to believe that it is a mystery into which the human mind will never penetrate."[6]

Euler's statement still echoes down through the centuries.

Here are two relatively simple puzzles concerning conjectures in number series that the recreational mathematician might enjoy tackling:

1. A young schoolgirl was playing with numbers one day when she performed the following calculations:

$$1^3 + 2^3 = 9$$

$$2^3 + 3^3 = 35$$

$$3^3 + 4^3 = 91$$

$$4^3 + 5^3 = 189$$

$$5^3 + 6^3 = 341$$

$$6^3 + 7^3 = 559$$

And so on.

The young girl noticed that all the numbers on the right-hand side of the equations were composite. She did many more similar calculations of summing two consecutive cubes, but each time, she found that the sum is always composite. She conjectured that this was always the case. The question is, is her conjecture correct?

2. A teacher presented a bright mathematics student with the following series: 1, 2, 3, 4, 5, 6, 7, 8, 9, 29, 34, 46, 57, 61. . . . The teacher asked the student to discover the next term in the series. The student succeeded in doing so.
   Can the reader discover the next term?

Solutions

1. The young girl's conjecture is correct. The sum of the cubes of two consecutive integers is $a^3 + b^3 = (a + b)(a^2 - ab + b^2)$. Therefore, the result always has at least two factors and is therefore composite.
2. The series consists of each number, n, such that n minus the product of the digits of n results in a palindromic number. Thus, the next number in the series after 61 is 78. The series then continues 82, 93, 101, 129, 143, 187, 202, 218, ...

CHAPTER 6

Does Mathematics Exist outside of Human Minds?

The question of whether mathematics is invented or discovered is an ancient one. Those who believe that human beings *discover* mathematics are known as *Platonists*. They adhere to Plato's view that mathematical objects have a mysterious reality outside of human minds. In their view, the number 17 is prime not because we define it as prime but because that is the way mathematical structure is built. The only factors of 17 are 1 and 17 itself.

Opposing this point of view are the *nominalists*. They believe that we *invent* mathematics. Hence, in their view, mathematics has no reality outside of human minds. Nominalists concede that we discovered the integers and that the relationships between the integers that followed that discovery were *invented*.

Consider the well-known number π. It is not only irrational but also transcendental. This means that π cannot be the root of any number. Its decimal expansion continues forever, without repeating. In 1974, the first million *decimal* digits of π were calculated in France. The one-millionth *decimal digit* was found to be 1. (If one includes the first 3 in the count, the one-millionth digit of π is 5.) The question now is this: *before* the one-millionth decimal digit of π was calculated, was the one-millionth decimal digit of π equal to 1? Or, put another way, did the one-millionth decimal digit of π exist *before* it was calculated in 1974?

Platonists, also known as *realists*, would answer yes to this question. They would argue that the one-millionth decimal digit of π has *always* existed, even before human beings evolved on this planet. They would say that every decimal digit of π exists, even the ones we have not yet calculated.

But the nominalists would strongly disagree with this conclusion. They say it is absurd to say that the unknown digits of π—those that we have not yet calculated—exist. No one knows the value of each of these digits. The nominalists would argue that only when a particular digit has been correctly calculated can one say that that digit of π *exists*. Several nominalists may concede that unknown digits of π *potentially* exist, but that is about as far as they will go on the subject.

The basis of mathematical Platonism may be put as follows: if nothing at all existed, it could be said (if there were someone there to say it) that the sum of all that exists is zero. The first axiom of set theory is that the *empty set* exists. The mathematical Platonist would say that the empty set has no elements. Therefore, its *cardinality* is zero. However, because the empty set exists, the total number of sets that exist is 1. Now, if that is true, there must be a second set holding the empty set. The cardinality of that second set is 1, for it contains the empty set. If that is true, there must exist a set that has two sets: the empty set and the other, second set, which has the empty set. Consequently, this means that there are three sets that exist. Of course, we can continue with this line of reasoning forever. Thus, each new set corresponds with a positive integer. First, we had one set: the empty set. The existence of the empty set means that a second set exists, which in turn means that a third set exists, and so on.

Why should we believe that mathematics exists outside of the human mind? To answer that, perhaps we should keep in mind that our best scientific theories could not be understood without the aid of mathematics. The indispensability of mathematics to making discoveries in science should give us good reason to believe that mathematics somehow exists *out there* in the world and is not merely a construct of the human mind.

Suppose you ask a computer programmer to use a computer-assisted drawing to construct a graph $y = 1/x$. Suppose the precise area under the curve from 1 to e is unknown to human beings. We assume that this is the first time that such a measurement is made. We also assume that no one has ever used a formula to calculate the area. The computer programmer does this work late in the evening. It is an easy matter for the computer user to use the computer to check the area under the curve from the point 1 to e, where e is the transcendental number 2.718281828459045. . . . However, let us assume the computer programmer retires for the night *without* checking what the area under the curve is.

The next morning, the computer programmer checks what that specified area is. She finds that the area is *exactly* equal to 1:

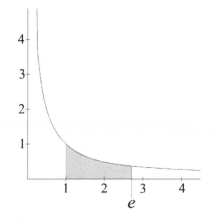

Graph of $y = 1/x$.

The question is this: Was the relevant area under the curve equal to 1 *before* the computer confirmed to the programmer that the area is indeed *exactly* equal to 1?

It seems ridiculous to suggest that the relevant numerical area under the curve did not *exist* until the first person discovered that the area is 1. The relevant area under the curve is inherent in the relevant graph. Given the stated graph, the specified area under the curve is *always* equal to 1 whether human beings have discovered that fact or not.

A similar argument can be made about any mathematical object. Consider the prime number 7. Before human beings evolved, was 7 a prime number? Mathematical Platonists would answer yes, 7 was a prime. They would go further. They would insist that the number 7 was *always* a prime and that it always will be a prime *whether human beings exist or not*.

The nominalists would strongly disagree with this view. They would argue that the number 7 (or any number) is a human invention. Their opinion is that most integers will have divisors besides themselves and 1. (Such numbers are called *composites*.) But the nominalists will be adamant that because of the invention of the integers, some integers will not have divisors except themselves and 1. (Such integers are called *primes*.)

In the opinion of several of the greatest mathematical minds on Earth, the number 7 is a prime number not because human beings say it is a prime but because *mathematical reality* is built that way. If we met aliens from space, they would undoubtedly have different symbols for the integers than the symbols we use. But whatever symbol they may have for, say, 7, they will tell us that the only integers that evenly divide into 7 are 7 itself and 1. In other words, like us, the aliens would have *discovered* that 7 is a prime number.

Let's refer again to the transcendental number known as e, which has a value of approximately 2.718281828459045. . . . It is believed that the Swiss mathematician Jacob Bernoulli (1655–1705) first discovered e in 1683. The number e is the limit of $(1 + 1/n)^n$ as n approaches infinity. That explains why e is ubiquitous in calculations involving compound interest. Later, it was discovered that in calculus, the number e is the only number equal to its own derivative. That explains why e is ubiquitous in calculus. If mathematics exists *only* in human minds, why is it not instantly recognized by human minds that e is equal to its own derivative?

When mathematicians discovered continued fractions, they found the following beautiful and elegant infinite continued fraction for e:

$$e = 2 + \cfrac{1}{1 + \cfrac{1}{2 + \cfrac{2}{3 + \cfrac{3}{\cdots}}}}$$

This continued fraction for e illustrates just how fundamental e is. If mathematics exists only in human minds, why is this continued fraction for e not at once clear to human beings? If mathematics exists only in human minds, why are the following two beautiful identities not at once obvious to human beings?

$$\text{limit as } n \text{ approaches infinity } n \times \left(\frac{\sqrt{2\pi n}}{n!}\right)^{\frac{1}{n}} = e$$

$$\text{limit as } n \text{ approaches infinity } \left(\frac{n!}{n^n}\right)^{\frac{1}{n}} = \frac{1}{e}$$

Of course, the nominalists will argue that the number known as e did not exist until it was *invented*. As far as nominalists are concerned, the following beautiful identity is an invention of human minds, not an eternal truth that human beings have *discovered*:

$$e^x = \frac{x^0}{0!} + \frac{x^1}{1!} + \frac{x^2}{2!} + \frac{x^3}{3!} + \frac{x^4}{4!} + \frac{x^5}{5!} + \frac{x^6}{6!} \cdots$$

If we let x equal 1, the above equation becomes

$$e = \frac{1}{0!} + \frac{1}{1!} + \frac{1}{2!} + \frac{1}{3!} + \frac{1}{4!} + \frac{1}{5!} + \frac{1}{6!} \cdots$$

Summing the first seven terms of this identity gives 2.718055. . . . One can easily see that the sum is rapidly approaching 2.718281828. . . , which is the value of e. If we make x equal 2 and sum the first seven terms, one obtains a value of 7.35555. . . . The sum of the infinite number of terms of this series is e^2, which is 7.38905. . . .

The nominalists will have us believe that the above identities are creations of the human mind and have no reality outside of human minds. However, if alien life exists elsewhere in the universe and has evolved to at least the level of intelligence that we have, they too will have discovered the astonishing number e and these beautiful equations.

Mathematical Platonists believe that these equations are eternal, timeless truths that intelligent beings *discover* and do not invent. As human beings continue to explore the structure of mathematics that is *out there*, new *discoveries* are made in various branches of its structure. The discovery of the number e was one of those fantastic discoveries. It concerns a fundamental and important constant in mathematics. The number e is now recognized as being ubiquitous in mathematics alongside 0, 1, π, and i.

Similar questions can, of course, be asked about the nature of π. Is π a construct of human minds? Or is π a number that we human beings have *discovered*? Consider a circle with a diameter of 1. The circumference equals π, which equals 3.14159265. . . . Like e, π can be expressed only as the sum of an infinite series or as an infinite continued fraction. The following beautiful infinite continued fraction equals π:

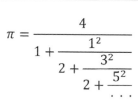

If mathematics is a human construct, why is this continued fraction for π not at once recognizable?

The following infinite series for π is known as Wallis's formula in honor of the English mathematician John Wallis (1616–1703), who discovered it in 1655:

$$\pi = 2 \times \frac{2 \times 2}{1 \times 3} \times \frac{4 \times 4}{3 \times 5} \times \frac{6 \times 6}{5 \times 7} \times \ldots$$

If mathematics is a human construct, why is Wallis's formula for π not at once recognizable?

The Scottish scientist James Clerk Maxwell (1831–1879) was one of the greatest theoretical physicists in history. He formulated a set of four beautiful equations that have become known as *Maxwell's equations*, which he first wrote down in 1865. His equations predicted waves of oscillating electrical and magnetic fields. These waves were eventually detected in a number of experiments by the German physicist Heinrich Hertz (1857–1894) in the late 1880s. These waves became known as *electromagnetic waves*. Maxwell's equations not only predicted electromagnetic waves but also predicted that they would travel at the speed of light. His equations led to the discovery of several things, including radio waves. Such is the remarkable power and beauty of mathematics. It can be justifiably argued that the electromagnetic waves were seen to be following mathematical laws that Maxwell *discovered*. We would not enjoy color television today or use our cell phones if it were not for Maxwell's discoveries.[1] Maxwell has been described as the greatest mathematical physicist since Newton.

In doing mathematics, it is difficult to believe that one is not discovering several aspects of reality. (Sometimes, in the media, one will read or hear of a new prime number that has been *discovered*. When did you ever read or hear that a new prime number was *invented*?) Consider the fact that there are an infinite number of solutions to finding a square integer that is the sum of two square integers. This is known as the Pythagorean theorem. This theorem exists for the following reasons. Choose any odd square integer greater than 1. Say you choose 9. Choose two consecutive integers that sum to 9. Because of the algebraic identity $(a^2 - b^2) = (a + b)(a - b)$, we can see that $(5^2 - 4^2) = (5 + 4)(5 - 4) = 9 = 3^2$. Therefore, $5^2 - 4^2 = 3^2$. Rearranging the terms, we obtain $3^2 + 4^2 = 5^2$. Thus, by choosing any odd square integer greater than 1, one will find that it can be expressed as the difference of two squares, thereby finding an infinite number of solutions of the sum of two square numbers equaling a square number.

No one, in any part of the world, has ever found that the sum of two cubes equals a cube or that the sum of two fourth powers equals a fourth power. The same goes for the sums of all higher powers. If aliens exist and they have discovered the integers, they will also find that $a^n + b^n = c^n$ exists in integers if n equals 1 or 2. But they will also find that such equations do not exist if the exponent is greater than 2. (In 1995, Andrew Wiles proved that the equation $a^n + b^n = c^n$ cannot exist in integers if n is greater than 2.[2])

The Pythagorean theorem not only is true in numbers, but also has a geometrical reality. On the plane, if one draws a right-angle triangle, one can prove that the sum of the squares on the two legs equals the sum of the square on the hypotenuse. However, no one in any part of the world can produce a drawing of two cubes whose volumes equal the volume of a third cube. Why? Because mathematical reality is not built that way. We cannot choose the laws of mathematical reality. We must accept mathematical reality for what it is. Mathematical reality imposes its laws on us, not the other way around.

In 2015, Carl Hagen (1937–), a particle physicist at the University of Rochester in New York, was teaching a class on quantum physics. He had introduced to his students a quantum mechanical technique known as the *variation principle*. This technique is used to *approximate* the energy states of quantum systems, like molecules, that can't be solved *exactly*. On the day in question, Hagen decided to teach his class how to use the technique in various atoms. Hagen then decided to teach his students how to use the technique to approximate the upper energy limit for each orbit that the electron makes of the proton in the hydrogen atom.

The hydrogen atom is one of the rare quantum systems in which the energy levels can be calculated exactly using conventional calculations. It is the simplest atom in the universe. In the hydrogen atom, one electron orbits one proton. (This has been the case for as long as hydrogen atoms have existed.) The electron occupies cloudlike orbitals of the proton. There is a distinct energy associated with each orbit. As Hagen compared the results obtained by the variation principle to exact results obtained by conventional calculations, he became intrigued by an unusual trend of the ratio of the approximate energies to the exact energies.

He called on his colleague Tamar Friedmann, a mathematician at the same university, to help him figure out what this trend was. They soon recognized that the mathematics involved in their inquiries was hiding Wallis's formula for π within it. Later, Hagen told the press that they had not been searching for π formulae. "The Wallis formula just fell into our laps," he said.

Kevin Knudson, professor of mathematics at the University of Florida and a math contributor for *Forbes* magazine, wrote that it all seemed like magic. He reportedly declared that the fact that a formula for π is hidden inside the quantum mechanics of the hydrogen atom is surprising and delightful.[3]

It seems reasonable to assume that those equations involving π that were found in the equations for the hydrogen atom have existed in some sense for as long as there have been hydrogen atoms. To say that those equations in which Wallis's formula makes an appearance were somehow *invented* appears nonsensical.

Several eminent scientists and mathematicians have long believed that mathematics exists *out there*. They do not believe that it is confined to the minds of human beings. As an example of this, consider again an electron orbiting the nucleus of an atom. Physicists believe electrons have a dual reality: they are both particles and waves. Physicists consider electrons as a cloud of probability density *and* simultaneously as a circular standing wave that surrounds the nucleus. They believe that any standing circular wave that is the electron must have an *integer* number of wavelengths to exist.[4] Thus, long before human beings walked Earth—indeed long before Earth was formed—electrons had to have an *integer* number of wavelengths. The only logical conclusion to draw from that fact is that integers have existed for at least as long as electrons have existed.

There are other examples to support the view that mathematics exists outside of human minds. The physicist Erwin Schrödinger (1887–1961) developed an equation in 1925 that became known as Schrödinger's equation. This equation is a partial differential equation that is used to obtain information on how an electron behaves in being bound to a nucleus. One of the surprising things about this equation is that it has i, which equals the $\sqrt{-1}$. Thus, an equation describing what happens at the level of the tiniest particles of matter includes the square root of negative 1.[5]

The English physicist Paul Dirac (1902–1984) had developed mathematical equations in 1929 that predicted that the *positron* exists. (The positron is an antimatter particle.) In 1929, there was no evidence that antimatter exists. However, in that year, Dirac announced that the positron exists simply because the mathematics implied that it exists. The English physicist had put his faith in mathematics. Because mathematics had shown that antimatter exists, Dirac believed that antimatter *did* exist. He once said, "One should allow oneself to be led in the direction which the mathematics suggests." That is what he precisely did in this instance.[6]

Dirac was later proved right for having placed his faith in mathematics, when three years later, in 1932, Carl David Anderson (1905–1991) discovered the existence of positrons in cosmic rays. Anderson later won the Nobel Prize in Physics in 1936 in recognition of his discovery.

In 1962 and 1963, researchers had shown that if an unusual type of *field* exists throughout the universe, it would result in several fundamental particles gaining mass. This field became known as the Higgs field, named after the British theoretical physicist Peter Higgs (1929–), who was one of the researchers. Although there was no known proof at the time that the Higgs field exists, the mathematics behind its hypothetical existence accurately described known particles at the time and predicted the existence of several other particles. During the 1970s, these predictions became known as the *Standard Model* of particle physics.

For decades, there was still no proof that the Higgs field exists. The technology needed to prove the existence of the Higgs field did not exist at that time. However, because of the successful predictions made by the Standard Model throughout the 1980s, scientists believed that the Higgs field was *likely* to exist. They also believed that if the Higgs field did exist, it would mean that a particle called the Higgs boson also exists.

The existence of a new subatomic particle with the expected properties of the Higgs boson was confirmed in collisions of subatomic particles in the Large Hadron Collider at the European Organization of Nuclear Research (also known as CERN) near Geneva, Switzerland, in 2012. Physicists now know that the Higgs field manifests itself through the existence of the Higgs boson. Physicists also believe that the Higgs boson is a scalar boson with *zero* spin.[7]

Furthermore, physicists believe that the excitations of a quantum field come in units called quanta.[8] (These quanta are perceived by humans as particles.) These units can be counted. When quantum fields interact, the number of excitations in a field can either increase or decrease *one unit at a time*. This seems to strongly imply that the integers have existed for as long as quantum fields have interacted, which is as far back as the Big Bang itself. This example profoundly illustrates that the integers are part of reality.

Of course, there are other ways to show that it is logical to believe that mathematics exists independently of human minds.

Consider the *Borsuk–Ulam theorem* in topology.[9] The theorem concerns itself with spheres and functions, but it can be applied to Earth's surface. The theorem was first proved in 1933 by the Polish mathematician Karol Borsuk (1905–1982). The first formulation of the theorem as we know it today was given by the Polish American scientist Stanislaw Ulam (1904–1984). Over the years, several other proofs of the theorem have been discovered. The theorem states that at any given moment on Earth's surface, there are two antipodal points (points that are on exactly opposite sides of the globe) that have the same temperature and barometric pressure.

The question that needs to be asked is this: Before the existence of human beings, were there two antipodal points on Earth's surface that had the same temperature and barometric pressure? It seems absurd to say that there were not. If human beings did not exist, the temperature and barometric pressure on two antipodal points of Earth would still be equal at any given moment. The only way to explain the Borsuk–Ulam theorem is by mathematics. That is because it *is* a mathematical theorem. That mathematical theorem was true long before human beings evolved on this planet. It will still be true when human beings no longer exist. It appears absurd to argue otherwise.

I conclude with one final example of how mathematics appears to have an existence outside of human minds. The German astronomer and mathematician Johannes Kepler (1571–1630) studied the periods of the planets and their distance from the sun and gave the following mathematical relationship, which is known as *Kepler's Third Law*: for every planet in our solar system, the ratio of their period squared to their semimajor axis cubed is the same constant value.

This means that $P^2/a^3 = 1$ if P is expressed in Earth years and if a is expressed in *astronomical units*. (An astronomical unit is defined as the mean distance between the sun and Earth. Its distance is approximately 93 million miles, or approximately 150 million kilometers.)

Thus, if we know either the period of the orbit or the mean distance of a planet from the sun expressed in astronomical units, we can find the other unknown value. For example, Uranus is about 19.19 astronomical units from the sun. To find how long it takes for Uranus to orbit the sun, cube 19.19, obtaining 7,066.83+. Then obtain the square root of 7,066.83+, obtaining 84.06. Thus, Uranus orbits the sun once in about 84 years. If you are told that Neptune orbits the sun once about every 164.8 years, how does one calculate the distance between the sun and Neptune? To do so, square 164.8, obtaining 27,159.04+. Then obtain the cube root of 27,159.04. The answer is about 30.05. Thus, Neptune is about 30 astronomical units from the sun.[10]

Briefly, Kepler's law tells us that before human beings ever evolved, the planets followed mathematical laws as they orbit the sun. They still follow mathematical laws to this day and will continue to do so into the future as they orbit the sun. To say that these mathematical laws did not exist for billions of years as the planets orbited the sun is, in the opinion of several great mathematicians (e.g., John Conway, Paul Erdős, and many others), absurd.

Finally, it is curious that modern scientists and non-Platonic mathematicians do not seem to have any difficulty in believing that several theoretical entities, such as electrons, protons, neutrons, photons, and quarks, all exist—independent of human minds—as part of the reality of the universe.

Why is it difficult for them to believe that mathematics also independently exists as part of reality and is not confined to human minds?

CHAPTER 7

Geometry Gems

When I was a schoolboy, geometry was a relatively popular branch in the school curriculum. However, over the years, geometry appears to have lost some of its appeal to students. Attempts have been made to reverse this trend. In 1960, Harold Coxeter published his famous book *Introduction to Geometry*. Seven years later, Coxeter followed up with a second important volume, *Geometry Revisited*. In more recent years, Alfred Posimentier published *The Secrets of Triangles*. All of these books are wonderful and educational in their own unique way.

These books contain many beautiful theorems in geometry. But many of those theorems are quite complex. In this chapter, I present the reader with several theorems that are relatively easy to understand but that are quite beautiful nevertheless.

Let us begin with probably one of the best-known theorems in mathematics: the Pythagorean theorem. Many adults recall this theorem from their schooldays. Consider any right triangle. Construct squares on each of its three sides. The square on the longest side is equal in area to the sum of the other two squares. It is a beautiful theorem but also an especially useful one. Carpenters, builders, and engineers all make use of this theorem in their day-to-day work.

There are an infinite number of Pythagorean right triangles. Pythagorean right triangles are triangles in which each of the three sides is an integer.

However, many who are familiar with this theorem are not familiar with this surprising fact: the inscribed radius of a Pythagorean right-angle triangle is an integer. The simplest Pythagorean right triangle is the 3, 4, 5 right triangle (see the figure at the top of page 46). The inscribed radius of such a triangle is 1. The second simplest Pythagorean right triangle is the 5, 12, 13 right triangle. The inscribed radius of such a triangle is 2. The third simplest Pythagorean right triangle is the 7, 24, 25 right triangle, whose inscribed radius is 3. And so on. An easy method to calculate the radius of a primitive Pythagorean right triangle is to divide the area of the triangle by half of the perimeter. Alternately, if the three sides of the right triangle are a, b, c, then $(a + b - c)/2$ equals the radius of the inscribed circle.

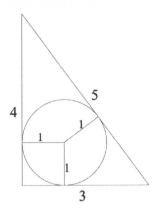

Another curious and beautiful result is Bottema's theorem. Beginning with any given triangle, construct squares on any two adjacent sides. Connect the vertices of the squares opposite the common vertex. Let M equal the midpoint of that line segment (see the figures below). Surprisingly, the location of M is independent of C.

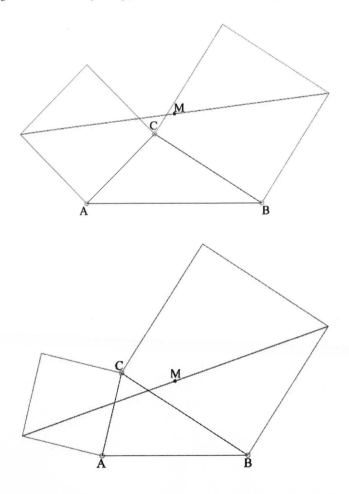

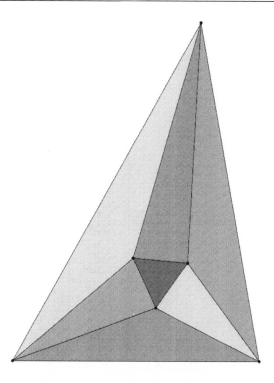

Morley's theorem is another beautiful result. Consider any triangle. Construct the three points of intersection of the adjacent trisectors of the triangle. These three points of intersection will form an equilateral triangle (see the figure above). This astonishing result was not discovered until 1899 by the English American mathematician Frank Morley (1860–1937).

The figure below illustrates another wonderful result. On the sides of any triangle, construct equilateral triangles, either all outward or all inward. Napoleon's theorem states that the lines connecting the centers of those equilateral triangles themselves form an equilateral triangle.

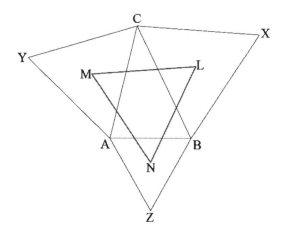

In 1878, a mathematics teacher named H. H. van Aubel (1830–1906) at the Royal Atheneum in Antwerp, Belgium, published a result that is now known as van Aubel's theorem (see the figure below). Squares are first constructed on the sides of any quadrilateral. Then two line segments are drawn from the centers of opposite squares. Van Aubel's theorem states that these two line segments are of equal length and are at right angles to each other.

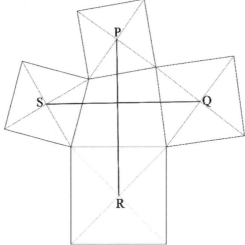

Finally, we mention one of the most beautiful and surprising relationships in plane geometry. Consider a random triangle that is illustrated in the figure below. Nine points are selected in the triangle as follows: the midpoints of the sides, the feet of the altitudes, and the midpoints of the segments from the orthocenter to the vertices. Astonishingly, all nine points lie on the same circle. This theorem applies to all triangles. This beautiful result is known as the nine-point circle theorem. The German geometer Karl Feuerbach (1800–1834) is often credited with discovering the nine-point circle theorem.

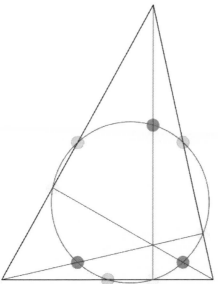

There are numerous problems in geometry that can be considered recreational. Theorems involving equilaterals are ideal for recreational puzzles. One of these is Viviani's theorem, named after Vincenzo Viviani (1622–1703) (see the figure below). Consider any interior point, P, in any equilateral triangle. Construct the three altitudes from P to the sides of the equilateral triangle. The theorem states that the sum of the three altitudes equals the length of the triangle's altitude.

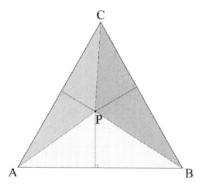

One interesting geometry problem goes as follows. Consider a point P in an equilateral triangle (see the figure below). Let P be three units from one vertex, four units from a second vertex, and five units from the third vertex. Determine the length of the side of the triangle.

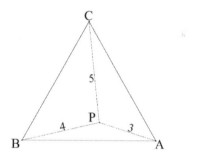

Solutions to this problem have appeared in various books over the years. Charles W. Trigg gives a beautiful solution in his book *Mathematical Quickies* (1967). The same solution is to be found in chapter 5 of Martin Gardner's book *Mathematical Circus* (1981).

I give a slightly different solution at the end of this chapter.

Solution

Let the side of the equilateral triangle equal d. Construct two triangles on the base of the equilateral triangle as shown in the figure at the top of page 50. There is now a smaller equilateral triangle that has a side length of 3, where each of the three angles is 60 degrees. One also has another triangle that has sides of 3, 4, 5. This triangle is therefore a right-angle triangle. Therefore, the angle BPA consists of a right angle plus

60 degrees, or 150 degrees. Consequently, we now know two of the sides and one angle opposite side *c*. Using the law of cosines, we can now determine the length of the side of the equilateral triangle.

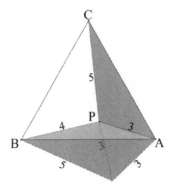

Thus

$$BA = \sqrt{a^2 + b^2 - 2\,ab \cos BPA}$$

$$BA = \sqrt{3^2 + 4^2 - 24 \cos 150}$$

This equals

$$BA = \sqrt{25 - 24 \times -0.866025}$$

This equals

$$BA = \sqrt{25 + 20.7846}$$

$$BA = \sqrt{45.7846}$$

$$BA = 6.76643$$

Thus, the side of the equilateral triangle equals 6.76643.

Other versions of the same problem are often asked. For example, the three distances given may be 5, 12, 13. (The three distances given must be such that they also form a triangle.)

When the three given distances are 5, 12, 13, the side of the equilateral triangle is 16.52038282.

The following beautiful equation holds, where the sides of the three distances are *a, b, c* and *d* is the length of the side of the equilateral triangle:

$$3(a^4 + b^4 + c^4 + d^4) = (a^2 + b^2 + c^2 + d^2)^2$$

Thus, if the lengths of the three distances are known, the equation can be used to determine d, which is the side of the equilateral triangle.

In our example, for a, b, c, we substitute 3, 4, 5. Applying this to our equation, we obtain

$$3(81 + 256 + 625 + d^4) = (9 + 16 + 25 + d^2)^2$$

This reduces to

$$193 + d^4 = 50d^2$$

This quartic equation can be changed to a much more manageable quadratic equation if we let $x = d^2$. The equation then can be written as $x^2 - 50x + 193 = 0$. Solving this quadratic, one finds that $x = 25 \pm 12\sqrt{3}$, which means that x equals 45.7846 or that x equals 4.2153.

The only solution to this quadratic equation that fits the problem's conditions is that x equals 45.7846. Since x equals d^2, we find that d equals the positive square root of 45.7846, or 6.766432568. Thus, the side of the equilateral triangle is 6.766432568.

If the three given distances are 5, 12, 13, the equation reduces to the following quartic: $17,761 + d^4 = 338d^2$. A similar trick to find the value of the side of the equilateral triangle can be used as previously. In this case, the only solution that fits the problem's conditions is that d equals 16.52038. Thus, the length of the side of the equilateral triangle is 16.52038.

The following formula will also determine the side of the equilateral triangle:

$$d^2 = \frac{1}{2}(a^2 + b^2 + c^2) + 2\sqrt{3}\,n, \text{where } n \text{ is the area of the triangle with sides } a, b, c.$$

If the three distances from P to the vertices of the equilateral triangle are a, b, c and the three sides a, b, c form a primitive Pythagorean right triangle in which the hypotenuse is one unit longer than the longer leg, the following pattern emerges:

$a, b, c,$	square of hypotenuse	twice the area of a, b, c	length of side of equilateral triangle
3, 4, 5,	25	12	$\sqrt{25 + 12\sqrt{3}} = 6.766432$
5, 12, 13,	169	60	$\sqrt{169 + 60\sqrt{3}} = 16.52038282$
7, 24, 25,	625	168	$\sqrt{625 + 168\sqrt{3}} = 30.26523642$
9, 40, 41,	1,681	360	$\sqrt{1681 + 360\sqrt{3}} = 48.00560687$
11, 60, 61,	3,721	660	$\sqrt{3721 + 660\sqrt{3}} = 69.7434838$

And so on.

If the length of the side of the equilateral triangle is *d*, the area of the equilateral triangle can be found by the following simple formula:

$$d^2 \times \frac{\sqrt{3}}{4}$$

We earlier found that if the three distances are 3, 4, 5, then *d* equals 6.76643. The area of the equilateral triangle is then easily found. It equals 19.825 square units.

If the three distances are 5, 12, 13, then *d* equals 16.5203. In that case, the area of the equilateral triangle is 118.179 square units.

If one is given *only* the three distances *a*, *b*, *c*, the area of the equilateral triangle can still be found. In such cases, the area of the equilateral triangle is

$$\frac{1}{2}\left[\frac{\sqrt{3}}{4}(a^2 + b^2 + c^2) \right] + 3 \; times \; the \; area \; of \; the \; triangle \; with \; sides \; a, b, c.$$

Thus, if the three distances are 3, 4, 5, the area of the equilateral triangle is

$$\frac{1}{2}\left[\frac{\sqrt{3}}{4}(3^2 + 4^2 + 5^2) + 3 \times 6 \right]$$

Solving, we find that the area of the equilateral triangle is 19.825 square units, as found previously.

CHAPTER 8

I First Meet with Dr. Moogle, Numerologist Extraordinaire

Although I am extremely interested in the phenomena of coincidence, I do not attribute their cause to supernatural sources. Nor do I believe in numerology. Thus, when a friend told me that a numerologist would be visiting Ireland for a week in November 2020, I showed no interest in the matter at all. But my friend was persistent, saying that this numerologist was extremely gifted at manipulating numbers.

"You will find him interesting," my friend said. "He is an unusual gentleman. His name is Dr. Moogle. He was an only child, born in India in 1978. His father was a Polish engineer, and his mother was a French teacher of mathematics. Dr. Moogle's father died young, and his mother and her son moved to France, where her son was reared. Dr. Moogle is reputed to be good company and extremely interested in recreational mathematics. For example, he told me that 2020 is equal to 16^2 plus 42^2. You might learn something from him. At the very least, you will find him amusing."

"I trust that Dr. Moogle is a sane character," I said. "I do not wish to meet with some unbalanced individual."

"He is quite sane, Owen," my friend said. "He is considered extremely intelligent. Dr. Moogle also has a good-looking daughter. She is aged about 25. She travels everywhere with her father. They seldom stay more than two weeks in any given place. I am sure you will find the father interesting. You might get some useful information that you can use in one of your books. I am sure you will find his daughter interesting also! Here is his cell phone number."

Two days later, I was on my way to meet Dr. Moogle in a house he had rented on the outskirts of Dublin. A stunning-looking lady in her mid-20s answered the front door when I called. She told me that her name is Anna. I quickly learned that she was Dr. Moogle's daughter. Anna directed me into her father's office. When I entered the dimly lit room, I saw a tall, smartly dressed gentleman with short, neatly cut hair sporting a meticulously kept and well-trimmed beard. I estimated that Dr. Moogle was about 45 years old. I approached him, and we shook hands.

"Mr. O'Shea," he said, as he spoke with a strong French accent. "I am pleased to meet you."

"I am pleased to meet you too, Dr. Moogle."

"I understand you are here to inquire about my work and my methods," he said.

"Yes, that's right. You can give me examples of your work. I may be able to use it, with your permission of course, in one of my books."

"Of course. Please take a seat. Would you like a coffee?"

"I would not say no to that offer," I replied.

"Good!" Dr. Moogle said. He pressed a button on his desk phone. "Anna," he said. "Would you be so kind as to bring two cups of coffee, sugar, and cream to my office, please?"

About five minutes later, Anna brought the coffee and chocolate biscuits into us. The coffee was nice, and the biscuits were delicious. It was difficult to take my eyes off Anna. She was an incredibly beautiful lady.

"Well, Mr. O'Shea," Dr. Moogle said, "I suspect you would like this interview to begin as soon as possible."

"That would be nice," I said. I took my notebook and pen from my inside jacket pocket.

"Okay, let's get down to business, shall we? First, let me point out that here we are in early November. Did you know that the Great War commenced in 1914? That is a number that is evenly divisible by 11."

"Is that significant?"

"It is when you realize that that terrible war ended at the 11th hour of the 11th day of the 11th month in 1918," Dr. Moogle said. "It appears that the numbers may have been telling us something."

"What were they telling us?"

"That we should notice so-called parallels more often and that by doing so we might avert some disasters. Let me give you a little example. As you may know, Owen, Belfast is known as the place where the RMS *Titanic* was built. A couple of similarities, if they had been noticed, may have forewarned that there was grave danger ahead for the *Titanic*."

"Please give me some details," I said.

"That famous ship was laid down in Harland and Wolff's shipyard in Belfast on March 31, 1909. That was the 90th day of '09. Note the mirror reflection of 90 and 09. One should be wary of such a significant date, especially when one is dealing with anything to do with water given the fact that water reflects just like glass! The date could have been interpreted as an ominous sign if it had been noticed. But apparently, no one did notice the similarity. On April 2, 1912, RMS *Titanic* sailed from Belfast to Southampton to begin its first voyage. It arrived in Southampton about midnight on the 4/4 and was towed to Berth 44. Apparently, no one noticed that numerical similarity either. If someone had done so, the alarm bells would have started ringing, and history would have taken a different course."

"Who is to say?" I said. "But I do accept your point. A slight change here and there can have such huge consequences in history."

"The Harland and Wolff shipyard number of the *Titanic*," Dr. Moogle said, "while she was under construction, was 401. I wish to show you a curious addition concerning the number 401 that I will come to in a moment. Let me first point out that the *Titanic* was laid down on March 31, 1909. She was launched on May 31, 1911. The *Titanic* sank on April 15, 1912. The survivors were brought to Pier 54 in New York City."

Dr. Moogle took a laminated card from a drawer in his desk and handed it to me. A tiny photograph of the *Titanic* was on the top left-hand corner of the card. The following addition sum was printed on the card:

A Coincidence Concerning RMS Titanic

By

Dr. Moogle

RMS Titanic was laid down on the 90th day of the year	90
RMS Titanic was launched on the 151st day of the year	151
RMS Titanic sank on the 106th day of the year	106
Survivors of the RMS Titanic were brought to Pier 54 in New York City	54
Total equals the Harland and Wolff shipyard number of RMS Titanic	401

I looked at the card. "Wonders will never cease!" I said. "May I hold on to the card?"

"Of course. Incidentally, the closest ship to the *Titanic* as she sank beneath the icy waters of the North Atlantic was the steamship RMS *Olympic*. She was the partnership of the *Titanic*. The *Olympic* was 401 plus 104 or 505 nautical miles from *Titanic* as she sank!"

"Gosh! That is surprising! I like the way the number 401 and its reverse both make an appearance in the distance between the two ships."

"I suspect," Dr. Moogle said, "that you did not spot that the number 505 equals $19^2 + 12^2$. The *Titanic* sank in 1912!"

"By heck," I said. "I never heard or read that before. It is certainly new to me. That is an eerie coincidence. How do you produce these curiosities?"

"I have a curious mind," Dr. Moogle said. "I look out for these kind of things."

"You are extremely good at spotting these events," I said. "I couldn't do it! Of course, many will say that all these curiosities can be ascribed to coincidence."

"I would not bet too many dollars on that," Dr. Moogle answered. "These types of coincidences happen much more often than the laws of probability predict. For example, I can give you, if you wish to hear them, many accounts of how numbers associate themselves with curious and famous events in history."

"Okay!" I said. "Please do!" At this stage, my appetite for unusual information had been whetted.

"Let me see," Dr. Moogle said. "Oh yes. Before I move on to other topics, please let me say a few more words about the *Titanic*. It hit the iceberg late in the day of April 14, 1912, and it sank to the bottom of the North Atlantic Ocean on the following morning, April 15. That date is often written as 4/15. I happen to like numbers, and I find the number 415 interesting." Dr. Moogle took a card from the drawer in his desk and wrote the following on it:

$$(4^5 + 1^5 + 5^5) \text{ divided by } (4 + 1 + 5) \text{ equals } 415$$

"That is an interesting number curio," I said. "It is completely new to me."

"I am glad you like it," Dr. Moogle said. "Incidentally, the number 505 is also interesting in this sense. It can be written as follows." Dr. Moogle took a sheet of paper from his desk and wrote the following on it:

$$505 = 10C5 + 10C0 + 10C5$$

I instantly checked that result on my scientific calculator. Dr. Moogle was, of course, correct.

"I can give you some more curious facts," Dr. Moogle said, "that have not been published before."

"Please do," I said, eagerly.

Dr. Moogle proceeded to give me some interesting information on the twentieth U.S. president, James A. Garfield. He said that President Garfield was born on November 19, 1831, and died on September 19, 1881, from gunshot wounds received in an assassination attempt on the president on July 2 of that year. Dr. Moogle told me that the first name of the medical doctor who treated President Garfield for his wounds was "Doctor." His full name, plus his title, is Doctor Doctor Willard Bliss. He was born in 1825 and died in 1889. Bliss grew up near to where James Garfield was reared, and the two appear to have known each other as youngsters.

Dr. Moogle also told me that palindrome numbers were strongly associated with James Garfield all his life. For example, Garfield was born on the 323rd day of the year. He married on November 11. He was promoted to the rank of brigadier general on January 11.

"I know," I said, "that on July 2, 1881, Charles J. Guiteau shot Garfield twice, with a .44-caliber Webley British Bulldog revolver. I notice that 44 is a palindrome number."

"Correct!" Dr. Moogle said. "President Garfield died from his wounds on 9/19/1881. The numbers 919 and 1881 are both palindromes. September 19 falls on the 262nd day of the year in a non–leap year. The number 262 is also a palindrome."

"That is amazing," I said.

"I am pleased you think so," Dr. Moogle said. "There is, you know, more than coincidence at play here. President Garfield was shot on 7/2. Using the usual alphabet code where A equals 1, B equals 2, and so on, the sum of the letters in the name

"James A. Garfield" adds up to 111. Add 72 to 111 to obtain 183. Curiously, President Garfield was shot on the 183rd day of the year."

"Astonishing," I said.

"Incidentally," Dr. Moogle continued, "the year number of Garfield's assassination, 1881, is a curious one. It equals the sum of four distinct positive cubes in two ways: 1^3 plus 3^3 plus 5^3 plus 12^3 equals 1,881, and 3^3 plus 5^3 plus 9^3 plus 10^3 also equals 1,881. Also, the following equation involving the prime factors of 1,881 is curious: 1,881 equals 3 times 3 times 11 times 19. The sum of the *digits* on both sides of the equation is equal."

"Marvelous!" I said, as I jotted the information in my notebook.

"In addition," Dr. Moogle said, "given the nature of this evil attack on Garfield, it is not surprising that the number of the beast, 666, was lurking in the background."

"How do you deduce that?" I asked. "The number 666 is nowhere to be seen in relation to the attack on Garfield."

"That is what you think, Mr. O'Shea! But I suggest you think again!"

Dr. Moogle then arose from his chair and walked to the large window at the side of the room.

"Important dates," he said, "especially ones involving assassinations, are not accidental, you know. Numbers have a life of their own, and they often make an appearance in notable events. The number 666 is no exception to this rule. Garfield was shot in 1881. Is it not strange that the number 666^{666} contains exactly 1,881 digits?"

"It certainly is," I said. "But consider the fact that President John F. Kennedy was assassinated on Friday, November 22, 1963. There is no appearance of 666 in that dreadful event?"

"On the contrary," Dr. Moogle said. "President Kennedy was shot on the *sixth* day of the week, by a shooter at a window, using the *sixth* floor of the Texas School Book Depository, while President Kennedy was one of *six* people sitting in the presidential car as it moved down Elm Street. You have three *sixes* here. Those three sixes can be combined to make the number 666."

"I would never have spotted that," I confessed.

"Here," Dr. Moogle said, "is something else that you probably would not have spotted. Consider the phrase 'Rifleman at Window.' Using the familiar alphabet code where A equals 1, B equals 2, and so on, the sum of the letters in the phrase 'Rifleman at Window' is 187. The media reported that Oswald was at the sixth-floor window of the Texas School Book Depository, when he shot Kennedy. If one multiplies 187 by 6, one obtains 1,122. Curiously, there are 187 primes less than 1,122. Kennedy was assassinated on 11/22. Here is a curious addition that is original with me that your readers may like."

Dr. Moogle walked back over to his desk. He sat down, opened a drawer in his desk, removed a card from the drawer, and handed it to me, on which the following was printed:

A Curious Addition Concerning the *JFK* Assassination

By

Dr. Moogle

John F. Kennedy was assassinated on the 326th day of the year	326
The room in which Oswald was questioned in by the Dallas police immediately after the assassination was numbered 317	317
The address of the Texas School Book Depository is 411 Elm Street	411
John F. Kennedy was shot in the 11th month	11
John F. Kennedy was shot on the 22nd day of the month	22
John F. Kennedy was the 35th US President	35
Total equals the date of the assassination 11/22	1122

I looked at the card. I must admit that I found the addition interesting.

"One decisive point about JFK," Dr. Moogle said. "The name of the street near Dealey Plaza in Dallas in which the Texas School Book Depository is located is 'ELM.' Using the familiar alphabet code, where A is 1, B is 2, and so on, the sum of the letters of ELM is 30. Of course, 30 is the newspaper reporter's traditional symbol for the end of a story. The street named 'ELM' was unfortunately the place where the story of the life of John F. Kennedy ended."

"But surely all this is coincidental," I said. "What about the late president's brother, Senator Robert Kennedy? Is there any appearance of 666 in his assassination in 1968?"

"There is," Dr. Moogle said. "When Robert Kennedy died, he was one of six family members in the room. He died on the sixth day of the sixth month. That accounts for the three sixes of 666."[1]

"I did not know that!"

"Numbers have always intrigued me," Dr. Moogle said. "When I was a small boy, when the school day was over, other boys played with toy soldiers. I always played with numbers. They became my lifelong friends! Numbers have given me so much joy in life! I am confident that I made the right choice! You know, I have always looked for the unusual and strange in life. Consequently, I have met many unusual and strange people over the years. I wonder if you are one of those. I will say no more about that at this moment in time. But I have always looked for the unusual in numbers also. For example, consider the present year: 2020. February 2, 2020, is a palindrome date when written in the month/day/year format or when written in the day/month/year format: 02-02-2020. Furthermore, that date fell on the 33rd day of the year, 333 days before the end of the year. Of course, 33 and 333 are both palindrome numbers."

"Wonderful!" I said as I scribbled this information into my notebook. "I am aware that the year 11/11/1111 is a palindrome year written in either of the date formats you mentioned. Was that the last time this phenomenon occurred?"

"Yes, it was. The next time this phenomenon will occur is on the 12/12/2121. The next such occurrence after that will be on 03/03/3030."

"I look forward to celebrating that day," I said.

"Good! Of course, you may be aware that the dates 4/4, 6/6, 8/8, 10/10, and 12/12 all fall on the same day of the week each year. The same applies for 5/9 (May 9) and 9/5 (September 5). All those dates fall on the same day each year! It was the late and great mathematician John Conway (1937–2020) who first spotted that curiosity."

"You are a mine of information," I said. "I should tell you that I am incredibly pleased to be receiving this very unusual information. Now, if you don't mind, let us get back to the present. This year, 2020, is a presidential election year. Can you tell me who is going to be elected the 46th president of the United States?"

"One can never be certain," Dr. Moogle said, "about predictions because the numbers *impel*; they never *compel*. But the numbers strongly suggest that Joe Biden will be chosen. You see, using the usual alphabet code where A equals 1, B equals 2, and so on, the sum of the values of the letters in the name JOE BIDEN equal 64, which is the reverse of 46. In addition, Biden was born on 11/20/1942. The sum of the digits of his birthday is 20, and the election is being held in the 20th year of the century. Also, the initials of his two names are the 10th and 2nd letters of the alphabet. Ten times two is 20. Thus, the numbers show that the number 20 will be significant in Biden's life. What could be more significant than being elected U.S. president in the year 20?"

"Indeed."

Dr. Moogle took a deep breath. "You may also like to know," he said, "that in Biden's lifetime (up to the day of inauguration as U.S. president), there are exactly 20 leap years."

"Amazing!" I said.

"Furthermore," Dr. Moogle said, "if Biden is elected, he will be sworn in as president of the United States of America on Wednesday, January 20, 2021. That date may be written as the seven-digit palindrome 1/20/2021. This is the first date in history that the inauguration day of a U.S. president occurs on a palindrome date. Two days after the inauguration day, on January 23, Joe Biden will be exactly 4,079 weeks old. The digits of that number sum to 20."

"By heck, you're right!" I said, "By the way, if Biden becomes president, he will be sworn in as president 78 days after the election, at the age of 78 years."

"Correct!" Dr. Moogle said, "the same age as you!"

"I'm not 78 years old."

"You could have fooled me," Dr. Moogle said. "In any event, did you know that if Biden wins the election on November 3, his first full day as U.S. president will be on the 21st day of the 21st year of the 21st century?"

"Holy mackerel! That's strange!"

"Yes, it is," the good doctor said. "A lot of curious things happen that cannot be accounted for by the laws of probability. I will give you a simple example of this. Did you know that the first handheld calculator was invented in Texas in 1967 by Texas Instruments? The year 1967 is a suitable year for the invention of a machine dealing with numbers given the fact that the 19th prime is 67. Of course, for a year intimately connected to numbers, it comes as no surprise that 1,967 equals $1,234 - 56 + 789$."

"Wow! I am certainly being educated today."

"In that case," Dr. Moogle said, "you should consider today to be a good day for you. Do you know of the brilliant singer Taylor Swift and how she considers 13 to be her lucky number?"

"I read that somewhere," I said. "Is it true?"

"Yes, it is true!" Dr. Moogle said. "Taylor Swift is a brilliant and world-famous popular singer. She was born on December 13, 1989. She turned 13 on a Friday the 13th. Her initials are the 20th and 19th letters of the English alphabet. Those two numbers sum to 39, which is three times 13. Her first record went gold in just 13 weeks. Her first number one song had a 13-second introduction. Taylor claims that every time she has won an award, she had been seated in either row 13, seat 13, section 13, or row M. Of course, M is the 13th letter of the English alphabet. Taylor's two albums *Fearless* and *1989* both have 13 songs."

"I admit that that is interesting," I said.

"The facts that I have just recounted about the number 13 in Swift's life are well known," Dr. Moogle said. "However, I will now tell you a couple of things about Taylor Swift's association with the number 13 that are *not* known, not even by Taylor Swift herself! These curiosities are original with me. Consider the singer's full name: Taylor Alison Swift. If one uses the usual alphabet code where A equals 1, B equals 2, and so on, the sum of the letters in the name 'Taylor Alison Swift' is 238. That number equals the sum of the first 13 primes. Plus, the sum of the digits in 238 is 13. Even the sum of the letters in the name 'Taylor A. Swift' equals 169, which is 13 squared."

"Holy mackerel!" I said. "How do you spot these things?"

"That is top secret," Dr. Moogle said. "Incidentally, the number 41 is not only the sum of the first 13 primes, but it is also the *13th* prime. Taylor Swift will celebrate her 52nd birthday on Friday, December 13, 2041."

"You are a mine of information, Dr. Moogle! I am certainly learning new things today! Of course, I repeat what I said earlier. Some people might look at all this and say that all these facts are mere number coincidences and that there is no substance to any of it. What message, if any, do you have for such people?"

Dr. Moogle shook his head.

"Mr. O'Shea," he said. "Please let me state this. I have no real interest in what people think of me or of my revelations. Such people, of course, are entitled to their opinions. But I would argue that I am also entitled to my viewpoint! I merely reiterate what I said earlier. There are far too many of these unusual and strange events involving numbers that can be explained by saying that they are mere coincidences."

"Perhaps you are right!" I said. "Who knows?"

Dr. Moogle nodded.

"There are strange events involving curious numbers," he said, "that simply cannot be explained by the laws of probability. I am sure you are familiar with the Tower of Hanoi puzzle. The traditional version of the puzzle usually consists of eight disks.

To solve this version of the Tower of Hanoi puzzle, a minimum of $2^8 - 1$, or 255, consecutive correct moves are needed. Here is a curious fact that is original with me. Using the usual alphabet code, where A equals 1, B equals 2, and so on, the sum of the letters in the words 'Tower of Hanoi Puzzle' equals 255!"

"By Jove! I have to admit that that is remarkable."

"Glad to hear you say that!" Dr. Moogle said. "It *is* remarkable! I made that discovery when I was just a boy of 10. It is an outstanding example of how unexpected numbers crop up in strange places. Let your readers know of that curious fact."

"I will."

"Why not let your readers contemplate this also? The U.S. Civil War has often been described as 'The War between Brothers.' Here's a curious fact. Using the alphabet code where A equals 1, B equals 2, C equals 3, and so on, the sum of the letters in the expression *United States Civil War* is 254. The sum of the letters in the expression 'The War between Brothers' also equals 254."

"Amazing!"

"Of course," Dr. Moogle said, "there are also curious similarities in the names of people and their occupations. There is a theory that one's name influences one's choice of occupation. The theory is known as *nominative determinism*. The word *aptronym* is often used to describe the phenomena. There is a massive amount of evidence to support the theory. For example, in the 1960s, there was a Chinese physicist named Kuan-Han Sun, who worked in Westinghouse Electric Corporation. He took part in research on solar radiation striking the moon. An article about Dr. Sun and his research, titled 'Dr. Sun, and the Moon,' was published in *Time* on Friday, October 28, 1966. The article ended by saying that Dr. Sun had made certain predictions about sunlight hitting the moon and that if those predictions prove to be correct, he would always be remembered as the sun who cast new light on the moon."

"That is funny," I said as I started laughing.

"Do you know of Tennys Sandgren? He is a well-known American athlete who plays tennis."

"You're joking!"

"No, I'm not," Dr. Moogle said. "It is a beautiful example of nominative determinism. There are countless other examples. Did you know that at one time, the head of the Department of Geology at the University of Western Australia was a man named Nick Rock? Or did you know that the ornithologist in charge of the Division of Birds at the Smithsonian's Museum of Natural History is Carla Dove? The tenth rector of the Church of the Ascension in New York City was Donald R. Goodness. Are you aware that at one time, the world 100-meter and 200-meter record holder was Usain Bolt? Is it not curious that Anton Horner was a member of the Philadelphia Orchestra for 44 years and its sole horn player for 28 years? The orchestra's musical director, Eugene Ormandy, described Mr. Horner as one of the greatest horn players of all time."[2]

"That is wonderful!"

"There are," Dr. Moogle continued, "two Republican politicians in the United States named John Doolittle and Tom DeLay. They have argued against any action being taken concerning the hole in the ozone layer."[3]

"Marvelous!" I said. "This curious phenomenon is also found with book authors. For example, in 1982, Bloomsbury Publishing PLC in London published *Advanced*

Motor Cycling. Its author is Geoff Carless. In 1968, the English publisher Books of Africa published *Geology of Southern Africa*. Its author is Edgar E. Mountain."

"Yes, it seems crazy, doesn't it," Dr. Moogle said. "I did some research on this topic a few years ago. I recall that in 1998, Oxford University Press in the United States published a book titled *Writing with Power*. Its author is Peter Elbow."

"Those authors' names are certainly curious," I said as I sat there on my chair, delighted to hear of these aptronyms. "I happen to like these curiosities. Many people are interested in this phenomenon. Did you know that the founders of the British Tarantula Society are Ann and Frank Webb?"[4]

"Yes, indeed," Dr. Moogle said. "I am aware of that. It all goes to show that we live in a strange world. How else can one explain your existence? Looking at you now brings to mind many strange things that I have read over the years. For example, there is a professor in the Department of Astronomy at Arizona State University named Sumner Starrfield. There was once a lawyer with the Sullivan and Cromwell law firm in New York City named Sue Yoo. In the 1990s, the British minister for agriculture, fisheries, and food was Douglas Hogg."

"It is strange!" I said. "But all this must be coincidental."

"I am not too sure," Dr. Moogle said. "There is something else responsible for this phenomenon. These strange occurrences happen far more often than the laws of probability would predict. For example, I recall reading a book titled *Pole Positions: The Polar Regions and the Future of the Planet*, published by Hodder and Stoughton in 1993. Its author is Daniel Snowman. I have also read that in England, the Royal Horticultural Society had a new president appointed in July 2020. His name is Keith Weed. It has been reported that Mr. Weed has had a lifelong love for gardening."[5]

"Wonderful!" I said. "I will give you one more example of this that I know of. *Did you know that at one time the Perth Airport executive general manager of corporate affairs and organization development was a lady named Fiona Lander?*"[6]

"It does not surprise me," Dr. Moogle said. "My dear Mr. O'Shea, we could talk about such matters all day."

Dr. Moogle got up, walked toward the window of the room, and looked out at the city of Dublin.

"I am aware," Dr. Moogle said, "that your books are published in the United States. Why not give this little piece to your American readers? It is a short little poem that I wrote when I was 12 years old. Your readers might like it."

Dr. Moogle handed me a sheet of paper that he took from his desktop. The page had the following lines:

AMERICA
America: I Think of Thee,
When I lay my head down to rest.
Land of my Fathers
My Dear Home in the West.
The stranger comes to your shore,
Seeking Peace and Prosperity.
Oh, America!
Dear Land of Liberty!

"That little poem," Dr. Moogle said, "has its own curiosities. Using the code where A equals 1, B equals 2, and so on, the value of the letters in the title, 'AMERICA,' sum to 50, which is the number of states in the USA. Using the same code, the sum of the letters in the little poem itself equals 1776; that is the year that the United States declared its independence."

"Brilliant!" I said. "Have you anything else?"

"Of course," Dr. Moogle said. "The following puzzle was passed on to me by Colm Mulcahy, professor emeritus at Spelman College in Atlanta, Georgia. You may have heard of him. He is well known in recreational mathematics circles. Consider this triangle."

He sketched a triangle on a sheet of paper as shown here:

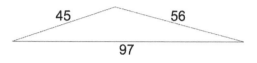

"Ask your readers to figure out what is so special about that triangle?"

"I will," I said. "Incidentally, my readers enjoy a good word puzzle now and then. Do you have any puzzle that I can give them?"

"Yes, I have. Here is a little cryptarithm puzzle."

Dr. Moogle took a small card from a drawer in his desk and handed it to me. The following words were written on the card:

```
    SIX
  SEVEN
  SEVEN
 TWENTY
```

"Each letter," he said, "in that addition problem stands for a different digit. No number begins with a leading zero. The problem is to find out what 'TWENTY' stands for. There is only one solution to this problem, but you will have to exercise a little gray matter to discover it."

"I will give it a try," I said. "Tomorrow, I hope to attend a lecture on elliptic curves. The talk is being given here in a local university, which I am looking forward to. But now I am afraid you will have to excuse me, as I must leave."

I placed my pen and notepad back into my jacket inside pocket.

I stood up to leave.

"Thank you, Dr. Moogle, for a remarkably interesting interview," I said as we shook hands. "I am sure I will be able to use some of your discoveries in one of my books. I look forward to meeting you again. May I have your cell phone number, please, before I go?"

"Of course," Dr. Moogle said. He scribbled his number on a scrap of paper and handed it to me.

"I will take your number too," Dr. Moogle said. I scribbled my number on a page of my notebook, tore out the page, and handed it to Dr. Moogle.

"Thank you, Mr. O'Shea. It is possible, I suppose, that I may get around to looking forward to reading your next book. Stranger things have happened! The American author Edgar Rice Burroughs (1875–1950) said, 'If you write one book, it may be bad. If you write 100, you have the odds in your favor.' Thus, try to get as many books as possible published!"

I left his room and walked into the hallway. As I did so, I paid a little more attention to his secretary, Anna. She was standing in the hallway, checking some files in one of the two filing cabinets that stood there by the wall. As I glanced at Anna, I wished that I were a young man again! She was a welcome sight! Anna had a perfect symmetrical face and long dark hair, which was perfectly matched by her short black skirt. As I went ahead to walk toward my car, my mind was full of curves—not the elliptic curves of number theory but the attractive physical curves belonging to Anna that I had been so lucky to have just seen.

Solution

1. The triangle, with sides equal to 45, 97, 56 that Dr. Moogle sketched, has an area equal to the square root of 459,756. One can use *Hero's formula* to calculate the area of a triangle if one is given the lengths of the three sides, a, b, c. Let the semi-perimeter equal s. Hero's formula then is

$$\sqrt{s(s-a)(s-b)(s-c)}$$

Substituting the values of our problem into the formula gives

$$\sqrt{99(99-45)(99-97)(99-56)} = \sqrt{459,756} = 678.053 = \text{area of triangle}$$

2. Dr. Moogle asked to substitute digits in the following cryptarithm to make the puzzle doubly true in letters and in numbers:

```
  SIX
SEVEN
SEVEN
TWENTY
```

The unique solution is

SIX	650
SEVEN	68782
SEVEN	68782
TWENTY	138214

CHAPTER 9

Simple Examples of How Mathematicians Derive Formulas

There are numerous formulas in mathematics that are used to solve myriad problems. How do mathematicians derive these formulas? These are questions that recreational mathematicians ask themselves from time to time.

Consider the formula for the summing of the partial series of integers from 1 to n. Suppose one wishes to find a formula to sum the first n integers. Here is one way to do it.

We let S equal the sum of the partial series, from 1 to n. Suppose we let n equal 6. We write this partial series down. We also write the terms of S in reverse:

$$S = 1 \quad + 2 + 3 + 4 + 5 + \ldots + n$$

$$S = 6 \ldots + 5 + 4 + 3 + 2 \quad + 1$$

$$2S = 6 + 1 + 6 + 1 + 6 + 1 + 6 + 1 + 6 + 1 + 6 + 1$$

We see that $2S = 6$ times $(6 + 1)$. Therefore, $S = (6 (6 + 1)/2)$.

Of course, we can let n equal any positive integer. Thus, the formula for the sum of the integers, beginning with 1, up to and including n, is $(n (n + 1)/2)$.

A little insight into the process that mathematicians use to derive various formulas can be gained by considering a mathematical process known as the *method of finite differences*.

The method of finite differences was first discovered by the English mathematician Brook Taylor (1685–1731). (He is the same gentleman for whom the Taylor series in calculus is named.) The Swiss mathematician Leonhard Euler (1707–1783) developed the ideas originated by Taylor. In 1860, George Boole (1815–1864) published several papers on Taylor's method.

Most readers can probably recall encountering quadratic equations in their school days. For those who recall these equations that have a strange sounding name, they will instantly recognize that quadratic equations are equations of the second degree. For example, $n^2 + 5n + 2$ is a quadratic equation. Quadratic equations are used to solve a variety of problems. Hence, they crop up quite often in mathematics.

Suppose one is given the following puzzle. What is the maximum number of pieces into which a pancake can be produced by n straight cuts, where each cut crosses all other cuts?

If no cuts are made, we have one piece: the pancake itself. One cut produces two pieces, two cuts produce four pieces, three cuts produce seven pieces, and four cuts yield 11 pieces. If we stop at this point, we have a series of numbers, 1, 2, 4, 7, and 11, being the number of pieces produced by the various cuts.

The problem now reduces to this: can we find a relatively simple formula that will produce this series of numbers? If we can, perhaps that formula can be used to find the number of pieces of the pancake that can be produced with five cuts, six cuts, and, in general, n cuts.

If the formula that we are looking for is not too complex, we can use the method of finite differences to find such a formula. Here is how we do it.

First, we write down a table like that shown below, where we form rows. Each row is the difference of adjacent terms in the row above. We continue to form these rows until we have a row where each term is similar. Thus, in the problem in our example, we obtain

Number of cuts of pancake	0	1	2	3	4
Number of pieces produced	1	2	4	7	11
First differences		1	2	3	4
Second differences			1	1	1

If the formula we seek generates a linear series, the first row of differences would have all similar numbers. If the formula we seek has a squared term, then the second row of differences would have all similar numbers. If the formula we seek has a cubed term, then the third row of differences would have all similar numbers and so on. If the formula we are searching for has, say, a term raised to the seventh degree, then the seventh row of differences will have all similar terms.

We see in our example that similar numbers appear in the second row of differences. This tells us that the equation we are looking for has a squared term, such as n^2.

Now that we have proved that the formula we look for is quadratic, how do we decide *exactly* what the formula is? There are several ways of doing this. Here is one method.

Consider the figure above. One-half of the term in the bottom row is the coefficient of n^2 in the formula. Thus, our formula has $1/2\ n^2$. The coefficient of n is equal to the first term in the middle row minus half the term in the bottom row. Thus, the coefficient of n is 1/2. The constant in the formula is the first number of the top row, which, in this case, is 1.

Thus, our formula is $1/2\ n^2 + 1/2\ n + 1$. Applying this to our puzzle about the pancake, we let n equal the number of cuts. When n equals 0, 1, 2, 3, 4, . . . , the number of pieces produced equals 1, 2, 4, 7, 11 (see the figure on page 67).

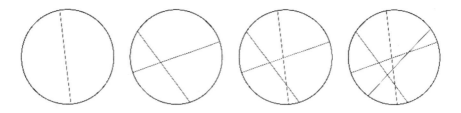

At this point, can we now say that we have found a correct formula for solving the pancake puzzle? Unfortunately, we can say only that "we probably have, but we cannot be certain." You see, in mathematics generally, we can find a formula (call it Formula X) for a precise reason that will give specific results in specific cases. Suppose we check the first four terms produced by Formula X. Is Formula X true in all cases? The answer is that it may be, but one must bear in mind that it may not be. Why? Because we must be aware that there are an infinite number of other formulas whose first four terms will correspond with those produced by Formula X, but whose terms thereafter differ.

We can be certain that Formula X gives correct results in *all* cases only after it has been mathematically *proved* that it does. If we use the formula we have derived, to find how many pieces are produced by five cuts, we obtain 16. If we try this in practice, we find that we can indeed cut the pancake into a maximum 16 pieces with five cuts. This strengthens our belief that the formula we have derived is correct in all cases. If we use the formula to find the maximum number of pieces the pancake can be cut into by six cuts, we obtain 22. Once again, if we try this in practice, we find that the pancake can be cut into a maximum 22 pieces with six cuts. This strengthens the belief even further that the formula we have derived is true in all cases. But it does not *prove* it.

We can prove that the formula we have found is true in *all* cases by mathematical reasoning as follows. When we cut the pancake, each nth line created by that cut crosses $n - 1$ lines. The $n - 1$ line divides the pancake into n regions. When the nth line crosses these n regions, it cuts each region into two parts. Consequently, every nth cut adds n pieces to the pancake. Before we begin cutting, there is one piece: the pancake itself. With the first cut, we add one extra piece. The second cut adds two more pieces. The third cut adds three more pieces, the fourth cut adds four more pieces, and so on. We find that the nth cut will add n more pieces.

Consequently, we find that the maximum number of pieces produced by n cuts is $(1) + 1 + 2 + 3 + 4 + \ldots + n$. Earlier, we found that the sum of $1 + 2 + 3 + 4 + \ldots + n$ is equal to $1/2\, n\,(n + 1)$. This can be rewritten as $1/2\, n^2 + 1/2\, n$. We must now add the first 1 in the series (the 1 in parentheses) to find the formula we are seeking. Thus, the formula for the maximum number of pieces is $1/2\, n^2 + 1/2\, n + 1$.

This is one of the ways in which mathematicians derive formulas when we are talking about relatively simple situations. However, this example will give readers an idea of how the mathematician derives formulas in general for finding the sum of various series.

Suppose a mathematician does not know the formula for summing the partial series of the first square integers: $1^2 + 2^2 + 3^2 + 4^2 + 5^2 + \ldots + n^2$. There are several ways she may approach the problem. Many of these methods are quite difficult and tedious to follow. However, here is one simple but ingenious method I found in George Pólya's *Mathematics and Plausible Reasoning* (2014).

First, calculate the first few sums of the squares of integers:

$$1^2 = 1^2$$
$$1^2 + 2^2 = 5$$
$$1^2 + 2^2 + 3^2 = 14$$
$$1^2 + 2^2 + 3^2 + 4^2 = 30$$
$$1^2 + 2^2 + 3^2 + 4^2 + 5^2 = 55$$

Now let the sum of the first consecutive integers $s(n)$ equal $1 + 2 + 3 + 4 + 5 + \ldots + n$.

Earlier in this chapter, we found that the formula for the sum of the integers, that is, $s(n)$, is $(n(n+1)/2)$.

Let $t(n)$ equal $1^2 + 2^2 + 3^2 + 4^2 + 5^2 + \ldots + n^2$. Construct the following table:

	1	2	3	4	5
First sums of squares $t(n)$	1	5	14	30	55
Triangular numbers $s(n)$	1	3	6	10	15

We are now going to find the ratio of each term in the sums of squares $t(n)$ to each corresponding term in the sum of the consecutive integers $s(n)$:

$$\text{Ratio } (1) = 1/1 = 3/3$$
$$\text{Ratio } (2) = 5/3$$
$$\text{Ratio } (3) = 14/6 = 7/3$$
$$\text{Ratio } (4) = 30/10 = 3 = 9/3$$
$$\text{Ratio } (5) = 55/15 = 11/3$$

We can see that the Ratio (n) appears to equal $((2n + 1)/3)$. (This conjecture can be proved by induction.)

For successive values of n, recall that Ratio $(n) = t(n)/s(n)$. Therefore, Ratio $(n) \times s(n) = t(n)$ for successive values of n. Thus, $((2n + 1)/3) \times (n(n+1)/2) = 1^2 + 2^2 + 3^2 + \ldots + n^2$. Or

$$\frac{2n+1}{3} \times \frac{n(n+1)}{2} = 1^2 + 2^2 + 3^2 + \ldots + n^2$$

Rearranging the terms of this equation, we obtain

$$\frac{n(n+1)(2n+1)}{6} = 1^2 + 2^2 + 3^2 + \ldots + n^2$$

This is the formula we look for.

Thus, we can use this formula to find, say, the sum of the first five squares. Here, n equals 5. The formula then is

$$\frac{5(6)\,(11)}{6} = 55 = 1^2 + 2^2 + 3^2 + 4^2 + 5^2$$

The triangular numbers often crop up in recreational mathematics. The formula for the nth triangular number is $1/2\,n(n+1)$. Beginning at 1, the triangular numbers are 1, 3, 6, 10, 15, 21, 28, 36, 45. . . .

Suppose a recreational mathematician wishes to sum the partial series of triangular numbers. This series is 1, 4, 10, 20, 35, 56, 84, 120, 165. . . .

Suppose the mathematician wishes to find a formula for the nth terms of this series. She can use the method of differences to find such a formula.

First, she writes the first few terms of the sequence. Then she continues to find the first, second, or third terms as follows until she reaches a row that has all similar terms:

Sequence	1	4	10	20	35	56	84	120
First differences		3	6	10	15	21	28	36
Second differences			3	4	5	6	7	8
Third differences				1	1	1	1	1

We see that the terms are all similar in the third row of differences. Therefore, the formula we look for is cubic. It must look like this: $U_n = an^3 + bn^2 + cn + d$.

Mathematicians have devised four equations to find the values of a, b, c, and d. These four equations can always be used when we are dealing with a cubic equation. These four equations are as follows:

$$6a = \text{third differences.}$$

$$12a + 2b = \text{first second difference.}$$

$$2a + 3b + c = U_2 = U_1.$$

$$a + b + c + d = U_1,$$

where U_1 is the first term of the sequence, U_2 is the second term, and so on.

Solving these equations gives the following values: $a = 1/6$, $b = 1/2$, $c = 1/3$, and $d = 0$. Therefore, the formula we seek is $U_n = 1/6n^3 + 1/2n^2 + 1/3n + 0$.

We can make this formula a little easier to work with by discarding the last term (0) and by multiplying across it by 6 and then dividing the entire formula by 6. The formula then becomes

$$\frac{n^3 + 3n^2 + 2n}{6}$$

This is the formula we look for. It will give us the sum of the first nth triangular numbers. Thus, if we wish to know, say, the sum of the first 10 triangular numbers, we substitute 10 for n in the formula. We then have

$$\frac{1000 + 300 + 20}{6} = 220$$

Thus, the sum of the first 10 triangular numbers is 220. This formula is true in all cases.

The method of finite differences does not always lead to a solution. For example, suppose we wish to find a formula that will produce the doubling series 1, 2, 4, 8, 16, 32. . . . If we use the method of finite differences on this series, we find that the first row of differences is 1, 2, 4, 8, 16, 32. . . . But this brings us back to where we started. Thus, in examples such as this, other means of finding a formula are needed.

Mathematicians have developed methods to find the sum of many series involving the integers. One of those general methods is *Faulhaber's formula*. It is named after the German mathematician Johann Faulhaber (1580–1635). Many details of Faulhaber's formula are quite technical and are probably outside the scope of what can be reasonably termed *recreational mathematics*. However, the following are several general results that are obtained from Faulhaber's formula that the reader may find interesting:

$$1 + 3 + 5 + \ldots + (2n - 1) = n^2$$

$$1^2 + 3^2 + 5^2 + \ldots + (2n - 1)^2 = \frac{(n)(2n - 1)(2n + 1)}{3}$$

$$2^2 + 4^2 + 6^2 + \ldots + n^2 = \frac{2n(n + 1)(2n + 1)}{3}$$

$$1^4 + 2^4 + 3^4 + \ldots + n^4 = \frac{n(n + 1)(2n + 1)(3n^2 + 3n - 1)}{30}$$

$$1^5 + 2^5 + 3^5 + \ldots + n^5 = \frac{2n^6 + 6n^5 + 5n^4 - n^2}{12}$$

CHAPTER 10

A Few Words about Simplicity, Mathematics, and Everything There Is

Readers of this book may be surprised to learn that there is an ongoing dispute between philosophers and scientists known as *regularists* and *necessitarians*. Regularists do not believe that there are laws of nature. They prefer to refer to these laws as statements of *regularities* in the world. For their part, necessitarians believe that there are laws of nature "out there" that describe how nature works. In their view, it is these laws that scientists try to *discover*.

Saying the same thing another way, regularists accept the world the way it is and believe that one cannot explain *why* the rules that govern the workings of the universe do so. Thus, while they accept the rules, they do not inquire any further into their origin. They believe that explanations must stop at some point. For them, that point is when one discovers the basic constituents of matter and the quantum fields that exist in the void. Their point of view is that the fundamental particles and the quantum fields are the rock bottom of reality. One cannot inquire any further. They believe it is pointless to ask, "Why these particles and why not other particles?" or "Why these rules of nature and why not other rules of nature?" The regularists accept the fundamental particles and the quantum fields as they are; they accept them as a brute fact.

On the other hand, necessitarians believe that there must be some reason the laws of nature are the way they are. Furthermore, necessitarians believe that it may have been possible that the laws of nature could have been different. Consequently, they ask, "Who or what chose the laws of nature of this universe?" Einstein once said that he wanted to know if God had a choice when he made the universe. Here, Einstein, who was an atheist, was using the metaphor *God* for the *laws* of *nature*. His question amounts to asking if the laws of nature could have been different.

Science is exceptionally good at answering "how" questions. How does an airplane fly from London to New York? The answer to this question—and similar "how" questions—is possible because there is a bigger explanatory context than the question itself. But science attempts to answer *only* science questions. It cannot answer metaphysics questions. If one asks, "Why is there something rather than nothing?," one is asking why a universe exists at all. This question, known as *the super-ultimate question*, is not a scientific question. It is a philosophical question. It is, in my opinion, unanswerable.

The super-ultimate question cannot be answered satisfactorily because there is no bigger explanatory context than the universe (or the multiverse). If we say that there are laws of nature that allow the universe to exist, one can ask, "Why do the laws of nature exist?" If scientists discover that there is a fundamental law, call it X, that manages all the laws of nature, one can ask, "Why X and not Y?"

In recent years, it has become almost fashionable to presume that the multiverse exists. *Multiverse* is the word usually used to describe an enormous assembly of (perhaps infinitely many) universes. The proposed existence of the multiverse is controversial. No evidence exists that confirms that the multiverse exists. However, many scientists and philosophers believe that there are solid grounds for believing in the existence of the multiverse. There are equally many scientists and philosophers who dismiss the idea as fantasy.

If one believes that the multiverse exists, in which the laws of nature vary from one universe to the next, one can argue from a statistical point of view that one can expect some universes to have laws of nature that will allow intelligent life to evolve. (Obviously, we must inhabit such a universe.) But one can then ask, "*Why* does the multiverse exist?"

At some point, the questioning cannot produce answers because there is *no* answer to the super-ultimate question of why there is something rather than nothing. If one says that quantum laws give rise to the universe, which apparently they do, one can ask where those quantum laws come from. Why do they exist? If one says that a supernatural being that one calls God manages the existence of the universe or multiverse, one can at once ask, "*Why* does God exist?" At the end of the questioning, the super-ultimate question, unanswered and unanswerable, is staring us in the face.

The mathematician explores the world of mathematics in a way equivalent to the way in which the physicist explores the laws of nature. The mathematician or physicist makes a discovery in their subject and believes that they may have detected a pattern. They make further investigations. If their findings are consistent, they make a hypothesis concerning their findings. As a starting point, they usually assume that the *simplest* hypothesis is the one that is most likely to be true.

Both the mathematician and the physicist have confidence that there is order in their chosen subjects. They may conduct their work without giving considerable thought to the philosophical implications of their inquiries. However, the physicist believes and works on the basis that there are laws of nature "out there" that can be discovered by human minds. She does not know where those laws of nature come from, but that does not deter her in her work. For his part, the mathematician believes or works on the basis that there is a mathematical structure "out there" that can be explored and believes that that exploration will sometimes (not always) yield solutions to the questions being asked of it.

Since Isaac Newton's (1642–1727) day, scientists, mathematicians, and philosophers have asked if the laws of nature are as simple as possible. It is strange, when one thinks about the matter, that the laws of nature are such that they allow for intelligent life to evolve on at least one planet in our universe. By doing so, that intelligent life then contemplates and investigates those very laws of nature that allow it to exist in the first place. It seems that intelligent life is nature's way of being aware of itself and of its laws, which gave rise to intelligent life. If intelligent life did not exist in the universe, those laws of nature would forever lay hidden and be unaware of themselves.

Of course, no one has been able to define exactly what they mean when they use the word *simple* when used in this context. A scientist who is a top-class mathematician may have one view on what constitutes simplicity in the laws of nature, but a colleague who is weak at mathematics may have a completely different view of what constitutes simplicity.

On one side of the debate (the non-simplistic view), there are many philosophers, scientists, and mathematicians who argue that there is no evidence to assume that nature is *simple*. On the contrary, they feel that the evidence shows that nature is, at its very core, complex. The adherents to this philosophy give numerous examples as to why they hold such a view. They point to the fact that in the past century, we have discovered that atoms, once believed to be indivisible entities consisting of a nucleus, are made up of protons and neutrons, which are surrounded by electrons. The electrons are fundamental particles, but the protons and neutrons are not. These protons and neutrons consist of even tinier particles that are named quarks.[1] It is believed that quarks are truly fundamental particles. Thus, they do not have any substructure.

In addition to all that, quantum mechanics has revealed that the world at the subatomic level behaves—to human observers—very strangely indeed. For example, quantum mechanics tells us that a particle can be both a wave and a particle.[2] At the subatomic level, the universe not only is strange but also appears to be stranger than we humans could ever have anticipated.

On the other side of the debate (the simplistic view), there are those who believe that there is an overarching simplicity in nature. They give numerous examples for holding the beliefs they hold. To support their point of view, they point to the fact that Newton deeply believed in the simplicity of nature's laws and that Newton's laws of gravitation were formulated on such a basis.

In Newton's day, it was known that the rate of acceleration of falling bodies at the surface of Earth is 32 feet per second every second. Newton knew from Galileo Galilei's (1564–1642) work that the distance that an accelerating body traveled was equal to $1/2\ at^2$, where a equals the acceleration in feet and t equals time in seconds. The equation states that a body at the surface of Earth will fall 16 feet in 1 second. Newton knew that the moon was about 60 Earth radii from Earth. By postulating an inverse square law, he calculated that the moon would fall a distance towards Earth in 1 second a distance equal to 1/3,600 of 16 feet. That works out about 1/20 of 1 inch. This result is correct.

Newton then went on to hypothesize that all the planets in the solar system followed an inverse square law. Thus, if the distance between two equally sized planets is doubled, the gravitational attraction between them is 1/4 of what it was initially. If the distance between the two planets is trebled, the gravitational attraction is 1/9 of what it was initially.

By careful calculations, Newton was able to use his theory to work out the densities of the sun, Jupiter, and Earth. This in turn led to the discovery that tides that formed in the oceans on Earth were caused by the gravitational pull of the moon. Newton was so confident of his discoveries that he named his theory the *universal law of gravitation*.

Of course, Newton could not *prove* that the force acting on planets as they orbited the sun was the same force that acted on apples on the surface of Earth. No mathematical equation could prove that result. All Newton could say was that *if* gravity exists and behaves the way his universal theory of gravitation predicted, it explains the movement of the planets in their orbit around the sun. In other words, Newton opted to choose the *simplest* explanation from several rival theories. He chose to accept that one theory—the one he called the *theory of gravitation*—managed the many varied phenomena that one can detect, ranging from falling bodies on Earth to the movement of the moon as it orbits Earth to the movement of the planets as they orbit the sun.

Newton's belief in the simplicity of nature resulted in a major advancement in understanding how the universe worked. Three hundred years after Newton formulated his theory of gravitation, his law proved to be so good and correct that NASA used it when sending men to the moon's surface.

The founder of the *theory of relativity*, Albert Einstein (1879–1955), also deeply believed in the simplicity of nature. His theory is a theory of gravitation, but it is much more exact, on a large scale, than Newton's theory of gravitation. Einstein is reported to have said that one should make things as simple as possible but not simpler. He also said that God (Einstein often used the metaphor of God when speaking of the laws of nature) would not give up the opportunity to make the universe as simple as possible.

Today, many philosophers and scientists (who hold the non-simplistic view) point out that Newton's law of gravitation, although correct over relatively small distances, is not as correct as the more complex and more difficult-to-understand theory of relativity. (Newton's law is said to be correct to 1 part in 10^7. Einstein's theory of relativity is said to be correct to 1 part in 10^{14}.)[3] Thus, in their view, this is strong evidence that nature's laws are not *simple*. They argue that all the evidence seems to support the view that nature's laws are complex.

For example, consider Einstein's famous equation concerning the equivalence of energy and matter: $E = mc^2$. The equation only consists of three variables, but each of these variables are found because of other not-so-easy-to-understand calculations that are needed to produce the values of the three variables.

Galileo first hypothesized that the orbits of planets around the sun were perfectly circular because the circle is such a simple geometrical shape. Later, Johannes Kepler (1571–1630) discovered his three laws, which showed that the orbit of a planet around the sun formed an ellipse. Is an ellipse more complex than a circle? The answer is debatable. However, it can be proved that elliptical orbits (rather than perfectly circular orbits) are by far the most likely paths a planet will take in its orbit. In other words, it is simply more probable that the orbit a planet takes will be elliptical.

Many scientists and philosophers to this day point out that this simplicity in nature is revealed by an enormous amount of basic physics formulas that involve relatively simple mathematical equations. Consequently, if there is much simplicity in nature, it seems to make sense to always try the simplest hypothesis first when trying to solve problems in physics.

This same reasoning applies also to one who is trying to solve a mathematical problem. For example, suppose you are a teacher teaching mathematics to children who have not yet learned the Pythagorean theorem. That theorem states that the sum of the areas of the squares on the two legs of a right-angle triangle *exactly* equals the area of the square on the hypotenuse.

You give a drawing like this to a bright child who is showing great promise in mathematics:

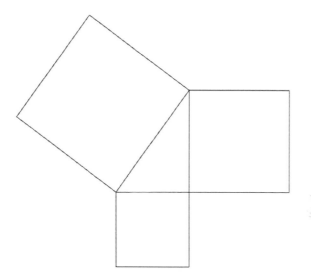

You ask the child if she can find any relationship between the three squares constructed on the three sides of the triangle. The child studies the drawing. Eventually, the child guesses the simplest explanation: that the joint areas of the squares on the two legs of the right-angle triangle *exactly* equal the area of the square on the hypotenuse. The child does not hypothesize that, say, 1.06 times the area of one square plus, say, 1.08 times the area of the other square exactly equals, say, 1.15 times the area of the square on the hypotenuse. No, the intelligent child does not go ahead in that manner. The child guesses the *simplest* hypothesis initially: that the joint areas of the two smaller squares equal the area of the largest square. The child will find that in this case, this hypothesis leads to the correct solution.

Now imagine a professional mathematician doodling with the aspects of a little-known number problem in mathematics. The mathematician does not know what the solution to the problem is. He initially plays around with the problem, trying this method first, then this other method, to find out if he can find a solution. The mathematician will almost certainly try first what appears to him to be the simplest hypothesis to get to the heart of the problem. All other things being equal, this procedure will more likely than not lead the mathematician to the correct solution.

Why should the simplest hypothesis prove so often to be the correct one? This is the question that has been asked by philosophers, scientists, and mathematicians since Newton's day. No one can answer this question definitively. It appears that the human mind seems to be programmed to believe that the simplest hypothesis is probably the correct one.

The reason for this may lie in our distant past when our ancestors first stood up and walked on the plains of Africa in search of food. If these ancestors saw what looked like a predator lurking in the tall grass and ready to pounce, they would hypothesize that it *was* a predator and not something that looked like a predator but wasn't. In such situations, our ancestors would most probably make their getaway quickly. They would survive and reproduce, passing on their genes and their ability to correctly see and escape from dangerous predators. The primitive human beings who were not able to spot danger were more likely to be killed, and therefore their genes would eventually be eliminated from the gene pool. The result of all this is that human beings alive today are obviously descended from those ancestors who were good at hypothesizing accurately. Consequently, human beings have a natural tendency to make a simple hypothesis on any given topic and believe that, often, that hypothesis is probably the correct one.

The question of whether nature is simple or complex can also be asked of mathematics. Mathematicians often surprisingly find that general results in mathematics can be more specifically and compactly expressed than they had first expected. For example, the French mathematician Pierre de Fermat (1601–1665) conjectured that the product of two prime numbers congruent to 3 or 7, modulo 20, can be expressed as $x^2 + 5y^2$. Fermat also conjectured that each such primes multiplied by 2 is also of the form $x^2 + 5y^2$. Later, the Italian mathematician Joseph-Louis Lagrange (1736–1813) and the French mathematician Adrien-Marie Legendre (1752–1833) proved both conjectures.[4] It seems that *sometimes* the structure of mathematics is such that otherwise complex results can be expressed in a simpler, more straightforward way. At other times, this simpler expression apparently cannot be achieved. Why this is so is a mystery that no one has ever been able to penetrate.

Stan Gudder (1937–), professor emeritus of mathematics at the University of Denver, said, "The essence of mathematics is not to make simple things complicated, but to make complicated things simple." Thus, the mathematician tries to formulate problems and solutions in the simplest manner possible. But in doing so, he must make himself aware that sometimes the apparently simplest result is not the correct one. Hence, he must go ahead with caution in his mathematical explorations and be ready to dismiss an apparently simple solution when it proves to be false.

Here is one example as to why it makes sense to distrust simplicity.

Suppose the bright child we mentioned earlier notices that the prime number 11 plus 2 equals another prime, 13. Then she notices that 13 plus 4 equals another prime, 17. She wonders if this is just a coincidence, so she investigates further. She does a little calculating and notices that 17 plus 6 equals 23. This is another prime. She begins to think that there is a pattern lurking here, one that somehow produces primes. She considers 23 and adds 8 to it, obtaining 31, which is yet another prime.

Now the young girl feels that she is on to something. She feels that there is some hitherto unknown pattern, connected to the distribution of prime numbers, hiding behind the simple integers and waiting to be discovered. She can almost see her name written up in lights because of her amazing discovery and the newspapers reporting that a mere child has discovered a pattern involving prime numbers that the experts have missed. The child takes 31 and adds 10, obtaining 41, which is also prime. Then she takes 41 and adds 12, obtaining 53, which is prime also. The she adds 53 and 14 together, obtaining 67, which is prime. She is really amazed at the results that she is getting. She adds 67 and 16 together, obtaining 83. Another prime. Next, she adds 83 and 18 together, obtaining 101. This is prime also. Then she adds 101 and 20 together. She is stunned with the answer. It is 121, which is *not* prime. It equals 11 times 11.

The pattern has broken down. The young girl's dream of discovering a pattern involving prime numbers is shattered.

This is what sometimes happens to scientists and professional mathematicians if they look for *only* simplicity in trying to solve one problem or another. The apparently simple solution is not always the correct one. Sometimes it is, but often it is not. Sometimes the solution to the problem is complex. This is the kernel of the matter. It seems that simplicity and complexity exist side by side in nature and in mathematics.

One example of where complexity and simplicity sit beside each other in mathematics concerns Legendre's conjecture about the distribution of the prime numbers. That conjecture has since been proved and is known today as the *prime number theorem*. The mathematics involved in trying to prove how the primes are distributed is complex. Letting x equal any given positive integer, Legendre's conjecture can be expressed as

$$\text{limit as } x \text{ approaches infinity of } \frac{x}{\left(\frac{x}{\text{natural log}(x)}\right)} = 1.08366 \ldots$$

I am unaware of Legendre's thinking process when he made his conjecture in 1798. It is believed that he had studied a list of the number of primes less than 1 million. He obviously did not have the list of primes that we have at our disposal today. But I suspect that the thought must have occurred to him that the limit of the above equation as x approaches infinity is *exactly* 1. This result would be so simple, elegant, and beautiful. Later, the Russian mathematician Pafnuty Chebyshev (1821–1894) proved that if a limit in the above equation exists at all, that limit *must* be 1. That limit is now known to be *exactly* 1. A proof of this was given by the Hungarian mathematician János Pintz (1950–).[5]

Here are some results of Legendre's conjecture. When x equals 1,000,000, the formula gives 1.08448. . . . When x equals 100,000,000, the formula gives 1.06129. . . . When x equals 1,000,000,000, the formula gives 1.05372. . . . When x equals 10,000,000,000, the formula gives 1.04779. . . . The limit *seems* to be approaching 1. The prime number theorem states that the limit is *exactly* 1.

Thus, we have a result, in advanced mathematics, that is beautifully simple and elegant.

We mentioned earlier Einstein's remarkable discovery that $E = mc^2$. Here, E stands for energy measured in joules, m stands for mass measured in kilograms, and c stands for the speed of light measured in meters per second. This equation illustrates that energy and matter are equivalent. In other words, energy and matter are just two sides of the one coin. Einstein's astonishing equation beautifully illustrates the power and simplicity that lie at the heart of nature. It helps us to calculate, for instance, that 1 gram of matter is equivalent to 21.481 kilotons of TNT. That is about the amount of energy released by the atomic bomb that destroyed Nagasaki in 1945.[6]

Ockham's razor is a well-known principle in science. It says that all other things being equal, the simplest hypothesis is usually the correct one. Several scientists believe it is a reasonable position to assume that the universe is as simple as it can be but sufficiently complex to allow life to begin and evolve to an intelligent level on at least one planet orbiting a nearby star.

Those scientists and philosophers who do not believe in the existence of the multiverse argue that the idea of a multiverse adds complexity to all that exists compared to the simplicity epitomized by the existence of one solitary universe.

However, if only one universe exists, it seems that there must be an additional law of nature, X, forbidding other universes from existing. On the other hand, if the multiverse exists, this added law of nature, X, forbidding a multitude of universes does not exist. In this sense, the existence of the multiverse is somehow *simpler* than if one solitary universe exists.

Proponents of the concept of a multiverse put forward the idea that if the quantum fields of mechanics can bring forth one universe from the vacuum, what is stopping the quantum fields in the vacuum from continuously bringing forth other universes, perhaps an infinite number of them?

It appears to be a reasonable idea.

Of course, there are those who contend that it is absurd to argue for the possible existence of a multitude of universes to explain the existence of the one solitary universe we inhabit. But they miss the crucial point. The vast majority of proponents of the multiverse argue for the multiverse theory not to explain this universe but because it appears to them to be the simplest hypothesis and therefore the most likely to be true rather than the hypothesis that there is only one universe.

There is, as said earlier, no scientific evidence at present to show that the multiverse exists. The whole concept of a multiverse is a very controversial subject, one that numerous physicists and cosmologists shun. They are unhappy that any concept outside of our universe, such as the idea of God or of the multiverse, is used to explain the existence of our universe.

But the belief in the multiverse may be entirely justified when one considers how immense nature is. Why would nature restrict itself to one solitary universe? At one time, human beings believed that the solar system is the entire universe. Subsequently, it was thought that the Milky Way galaxy is the entire universe. We then learned that there are billions of other galaxies in the universe. In other words, humankind discovered gradually that its first beliefs of what constituted reality were completely off the mark (to say the very least).

The proponents of the theory that only one universe exists complain that talk of the multiverse is nonscientific and mere speculation. But there have been areas of science that were once considered speculative and have turned out to be true afterwards.

For example, in 1928, the great English physicist Paul Dirac (1902–1984) postulated that antimatter exists in our universe. He based his opinion on the fact that the mathematics he was working on seemed to show that a particle with the same mass as an electron but with the opposite charge must exist. He called this new particle the *positron*. If the positron exists, then it means that antimatter exists. The scientific community was skeptical about Dirac's claim. But in 1932, Dirac's faith in the power of mathematics was vindicated when the positron was discovered. Contrary to what the scientific community believed at the time, antimatter *does* exist.

There are other similar examples. In 1916, Albert Einstein's theory of general relativity predicted the existence of *black holes*. These objects are so dense that they warp the fabric of space-time so much that even light cannot escape their gravitational attraction. At the time that Einstein's theory of general relativity had predicted black holes, none had ever been discovered, and many scientists were skeptical about their supposed existence. But in 1971, black holes were discovered. They do exist.

The theory of general relativity also predicted the existence of gravitational waves. They could not be detected in Einstein's day. Gravitational waves are ripples in the fabric of space-time caused by accelerating masses and other violent events in space-time. It took decades for scientists to develop the technology to conduct a search for them. It was only in 2016 that the Laser Interferometer Gravitational-Wave Observatory (LIGO) first detected them, proving their existence. In the following year, 2017, the Nobel Prize in Physics was awarded to three American physicists—Rainer Weiss (1932–), Kip Thorne (1940–), and Barry C. Barish (1936–)—for their contributions to the LIGO detector, which led to the observation of gravitational waves.

The reasoning used to suggest that the theory of the multiverse is somehow *simpler* than the theory that only one universe exists can be illustrated by considering the complexity of computer programs. Consider the shortest computer program that is used to produce a specific object as output. The Kolmogorov complexity is a measure of the computational resources needed to specify the object. (This measure of *algorithmic complexity* is named after the Soviet mathematician Andrey Kolmogorov [1903–1987], who first published on the subject in 1963.)

The well-known Swedish multiverse proponent Max Tegmark (1967–) points out that a computer program written to give as output the set of *all* integers can be generated by a relatively short program but that a computer program written to produce one *specific* integer can be far longer. From this perspective, one may argue that the program that generates the list of *all* integers is somehow *simpler* than the second program that generates only *one* integer. In other words, sometimes the production of a *collection* of specific things is simpler than the production of *one* specific thing. It seems that sometimes the creation of an ensemble can be simpler than the creation of a solitary object.

A universe that is as simple as possible but sufficiently complex to allow living organisms to evolve on at least one planet orbiting a nearby star does not contradict any known laws of nature. Our universe, it appears, may be one of the numerous different universes that nature may produce in the vacuum. In other words, our universe may be like one grain of sand on an infinite shore. A multiverse, having billions of universes, perhaps an infinite number of universes, is, in the view of many eminent physicists, most likely.[7]

Whether there is just one universe or a multitude, the super-ultimate question of why there is something rather nothing is still there, lurking in the shadows. As we inquire ever deeper into the origin of the universe, the super-ultimate question is like a brick wall that we cannot penetrate.

Several physicists state that the universe or multiverse has *always* existed. They usually back this statement up by saying that *nothingness* is unstable and give this as the reason for why there is something rather than nothing. These physicists (and several philosophers) believe that if this is the case, there is no further explanation for the origin of matter, space, and time.

But that conclusion cannot be correct. The super-ultimate question is still there. One can at once ask, why is it that *nothingness* is unstable? Who or what decided that to be the case?

Several physicists believe that it is only theists who use the super-ultimate question to defend theism. But numerous atheists (e.g., David Z. Albert) are deeply troubled by the mystery of existence. Those atheists obviously do not believe that the origin of the universe has a supernatural explanation. On the contrary, the majority of them believe that it has a *natural* explanation. There are other atheists, of course, who believe that the problem of existence is not susceptible to *any* solution. For them, there is *no* explanation possible as to why the universe exists.

There are physicists who argue that the universe came from nothing. The well-known American physicist Lawrence M. Krauss (1954–) authored a book about this idea titled *A Universe from Nothing: Why There Is Something Rather Than Nothing* (2012).

The English theoretical physicist and cosmologist Stephen Hawking (1942–2018) and the American physicist and mathematician Leonard Mlodinow (1954–) also argued that the universe could have come from nothing. In their book *The Grand Design* (2010), both physicists insist that because there is a law called gravity, the universe can create itself from nothing.

But the *nothing* described by Krauss, Hawking, or Mlodinow, from which the Big Bang appeared, is actually not *nothing*. It is *something*—a quantum field, running to quantum laws, which is capable of making at least one universe. Thus, the theory that Krauss (and Hawking and Mlodinow and others) propose, that the universe popped into existence from nothing, is fatally flawed. The universe popped into existence from something. This something consists of quantum fields. The super-ultimate question then reduces to this: Why do the quantum fields exist?

Is there something, call it *Z*, behind the quantum fields that exists from all eternity that "tells" the quantum fields to produce a universe? We do not know. If there is, the question of why *Z* exists immediately arises. Perhaps there is nothing at all behind the quantum laws. The quantum fields and laws are just there in the void, existing from all eternity. If that is the case, then somehow the quantum fields in the void just "know" that they are fertile and capable of producing at least one universe, and, ipso facto, a universe bursts forth in a Big Bang.

At the end of the day, we can say that the universe (or multiverse) exists because of the laws of quantum fields. But then we can ask, *Why* do the laws of quantum fields exist? Why are there quantum laws at all rather than completely nothing? Would not everything be much simpler if *nothing at all* existed from all eternity? For some reason, nature, it appears, has decided that the laws of quantum fields eternally exist.

Why is this?

No one knows.

No matter what the approach, the super-ultimate question is still there. It is still the deepest unanswered and unanswerable philosophical question that any intelligent being can ask.

Finally, here are nine examples of how mathematical calculations that one might expect to be complicated transform into something remarkably simple:

1. Consider the following expression, which the great Indian mathematical genius, Srinivasa Ramanujan (1887–1920), discovered, where, a, b, c, are positive integers:

$$(a + 1)(b + 1)(c + 1) + (a - 1)(b - 1)(c - 1) = 2(a + b + c + abc)$$

Note how the apparently complicated left-hand side of the equation simplifies to the relatively simple expression shown on the right-hand side.

2. The following is another Ramanujan expression, illustrating that it is possible for the equation $A^3 + B^3 = C^2$ to exist in integers.

$$(-3s^4 + 6t^2s^2 + t^4)^3 + (3s^4 + 6t^2s^2 - t^4)^3 = (6st(3s^4 + t^4))^2$$

In the equation above, s and t are chosen so that they are coprime of opposite parity, and t is not a multiple of 3.

Note how the sum of the two cubes on the left-hand side of the equation simplify to the squared expression on the right-hand side.

3. For all positive integers, n, the expression

$$\left(\frac{n^5}{5} + \frac{n^3}{3} + \frac{7n}{15}\right)$$

is an integer.

4. The largest possible sphere that can be inserted in a cube has a volume in relation to the volume of the cube equal to $\pi/6$.

5. The incircle of a primitive Pythagorean right triangle is always an integer. To obtain the radius of the incircle of such a right triangle, add the length of the two legs and subtract the length of the hypotenuse. Divide that result by 2, and you have the radius of the incircle.

6. Suppose one is given the length, x, of the short side of a primitive Pythagorean right triangle, where the hypotenuse is one unit longer than the longer leg. To find the perimeter of such a triangle, multiply x by $(x + 1)$. The answer equals the perimeter.

7. Consider the product of the following three complex numbers, where i is the square root of negative 1: $1 + i$, $1 + 2i$, and $1 + 3i$. Surprisingly, the product equals -10. That also equals the product of the following three complex numbers: $1 - i$, $1 - 2i$, and $1 - 3i$.

$$(1 + i) \times (1 + 2i) \times (1 + 3i) = -10 = (1 - i) \times (1 - 2i) \times (1 - 3i)$$

8. $(a + b + c)^2 + (a + b - c)^2 + (a - b + c)^2 + (a - b - c)^2 = 4(a^2 + b^2 + c^2)$

9. $\left(\frac{1}{2}\right)^3 + \left(\frac{2}{3}\right)^3 + \left(\frac{5}{6}\right)^3 = 1$

CHAPTER 11

Tom the Piper's Son and Other Classic Calculus Problems That Can Be Solved without Calculus

Most recreational mathematics books do not contain calculus problems. Presumably, this is because many of the readers of these books, although highly intelligent, may not have studied mathematics in college.

However, it is surprising that many mathematical puzzles, which usually require calculus to solve, often yield to some relatively easy formula or rule. The formula produces the correct solution, although calculus is needed to prove why the formula or rule works.

For example, consider the following problem, versions of which have appeared in countless introductory calculus books over the years. Suppose a plumber has a square sheet of zinc measuring 12 feet by 12 feet. He wants to cut off four square corners so that he can fold the sides up to make a cistern without a top. How large should the corners be if he wants to make an open-topped cistern that has the maximum volume?

Surprisingly, to find the size of the corners, simply divide the length of the square by 6. (One always divides the length of the square by 6.) In our example, the sheet of zinc measures 12 feet by 12 feet. Therefore, the size of the corners to be cut off equals 2 feet in length. Consequently, the cistern will measure 8 feet by 8 feet and will have a depth of 2 feet. Therefore, its volume is 128 cubic feet. This is the maximum volume that can be achieved. If instead one had made the size of the corners, say, 1.9 feet, the cistern's volume would be 127.756 cubic feet. If the size of the corners had been 2.1 feet, the cistern's volume would have been 127.764. Many recreational math enthusiasts find it extraordinary that such a simple formula will solve such a problem.

A second simple formula that will give the maximum volume of the box in cubic inches can be used when one is intending to make a box with maximum volume from a square sheet. Let a equal the length of the sheet in inches of metal. Then $2a^3/27$ will give the maximum height of the box in cubic inches. Applying this formula to our square sheet of metal measuring 12 feet, or 144 inches, on its side gives $(2 \times 2,985,984)/27$, or 221,184 cubic inches. Divide this by 1,728 to convert to cubic feet. The answer is 128 cubic feet. That is the maximum volume of the box that can be made from the square sheet measuring 12 by 12 feet.

The formulas just given apply to square sheets of metal. They are derived from a similar problem concerning a rectangular sheet of metal. For example, suppose one has a rectangular sheet of zinc that measures 8 feet by 3 feet (see the figure below). Once again, one wishes to cut off four square corners of the sheet and then fold up the sides and make an open-topped box. How large should the corners be if one wishes to make an open-topped cistern containing maximum volume?

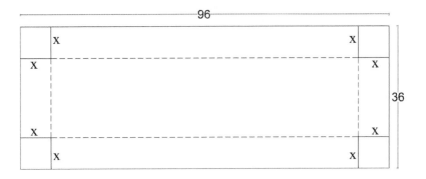

This question was posed by the English puzzle genius Henry Ernest Dudeney (problem number 201) in his book *Amusements in Mathematics*, first published in 1917. Dudeney solves this puzzle using a surprisingly simple formula. The simple formula can be used to solve all puzzles of this type.

If a and b are the sides of the sheet of metal, the following formula gives the sizes of each of the corners that will produce a cistern with maximum volume:

$$\frac{a + b - \sqrt{a^2 + b^2 - ab}}{6}$$

In our example, the sheet of metal measures 8 feet in length by 3 feet in width. The formula gives the size of the corners as 4/6 feet, or 8 inches. Therefore, the dimensions of the cistern are $(96 - 16)$ inches by $(36 - 16)$ inches by 8 inches. This gives a volume of 80 by 20 by 8, or 12,800 cubic inches. Dividing by 1,728 to convert to cubic feet gives 7.407 cubic feet as the maximum volume of the cistern.

If the sizes of the corner squares were 7 inches long, the resulting cistern would have a maximum volume of 82 by 22 by 7, or 12,628 cubic inches. This equals 7.307 cubic feet. On the other hand, if the sizes of the corner squares to be cut away were 9 inches long, the resulting cistern would be 78 by 18 by 9, or 12,636 cubic inches. This is equivalent to 7.3125 cubic feet.

That such a simple formula given by Dudeney can give the maximum volume of an open-topped cistern is remarkable.

The problem that Dudeney posed and solved by using a simple formula is approached using differential calculus as follows. Let x equal the length of each of the four corner squares that are cut away from the sheet of metal that measures 8 feet by 3 feet. The volume of the resulting box can then be expressed as $(8 - 2x)(3 - 2x)x$.

This equals $(24 - 6x - 16x + 4x^2)x$. This in turn equals $4x^3 - 22x^2 + 24x$. This is the function we want to maximize. Thus, we obtain the derivative of this function with respect to x and equate to zero:

$$\frac{dV}{dx} = 12x^2 - 44x + 24 = 0$$

This a quadratic equation. Using the quadratic formula to solve this equation gives x a value of 2/3 or x a value of 3. These are known as the *critical values*. If x equals 3, it will make the width of the box equal to $(3 - 6)$, or −3. That is impossible. We cannot have a box with negative width. Thus, x must equal 2/3.

But is this critical value a local maximum or minimum value? To find out, we will use the second derivative test for $x = 2/3$:

$$\frac{d^2V}{dx^2} = 24x - 44 = 4(6x - 11)$$

When $x = 2/3$, we obtain

$$\frac{d^2V}{dx^2} = 24x - 44 = 4(6x - 11) = 4\left(\frac{12}{3} - 11\right) = 4(4 - 11) = -28$$

Hence, since the second derivative is less than zero, $x = 2/3$ is a local *maximum value*. Before cutting out the four corners of the sheet of metal, the sheet measured 24/3 by 9/3. After cutting out the four corners and folding up the sides to make an open-topped box, the resulting box has a volume equal to $20/3 \times 5/3 \times 2/3$, or 200/27 cubic feet. This equals 7.407 cubic feet. This is the maximum volume possible given the dimensions of the sheet of metal.

Readers who are familiar with the calculus (and even those who are not) will appreciate how much easier the problem is solved by using the simple formula provided by Dudeney. Of course, Dudeney's formula gives the solution to the problem, but calculus is still required to prove that the result is indeed the maximum volume.

There are many calculus problems that can be solved without using calculus. Here is one where a little insight yields a solution. Consider the drawing below. One is given the dimensions between A and B and the distance between A and the edge of the desert. A company must pump water through a pipe from point A to point C on the edge of a desert and then on to point B. Where should point C be located at the edge of the desert that will provide the shortest length of pipe to be used? A little insight will show that a straight line connecting point A' to B through point C is the shortest distance between A' and B. Thus, point C is the point on the edge of the desert that gives the shortest pipeline.

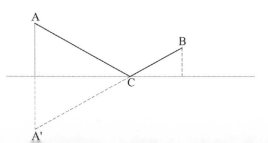

Consider the puzzle made famous by Sam Loyd, titled "Tom the Piper's Son." It appeared on page 217 of his *Cyclopedia of Puzzles* (1914). According to the old Mother Goose rhyme, Tom stole a pig. According to Loyd, Tom entered the field through the far gate on the right-hand side of the field (see below). The pig was rooting at the base of the tree on the far left-hand side. At this point, the pig was exactly 250 yards from Tom. Tom chased the pig as it ran directly down toward the left-hand bottom corner of the field. Both Tom and the pig ran at constant speeds. At all times, Tom ran directly toward the pig. It transpired that Tom could run one-third faster than the pig. The problem is to determine how far the pig ran before he was caught.

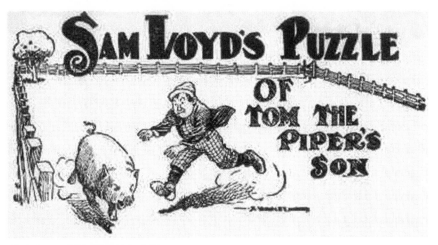

Courtesy of the Sam Loyd Company

This puzzle can be solved by calculus. However, Loyd gives a simple rule on page 367 of the *Cyclopedia* that can be used to solve the problem no matter what the initial distance the pig is from Tom. Furthermore, the rule given by Loyd shows that Tom will always catch the pig provided that Tom is traveling faster than the pig.

Here is Loyd's rule. First, calculate the distance Tom would have to travel if both Tom and the pig ran in a straight line in the same direction. Then calculate the distance Tom would have to travel if both he and the pig traveled in a straight line toward each other. Add the two results and then obtain half of that answer. The result is the total distance Tom would have to travel to catch the pig.

Let's see how this all works out in our problem. Since Tom travels 4/3 faster than the pig, if both were running in the same direction, Tom would catch the pig after running 1,000 yards. If both Tom and the pig were running toward each other, Tom would run 4/7 of 250 yards. This equals 142.857142 yards. We now add the two distances that Tom ran: 1,000 + 142.857142 = 1,142.857142. We now divide this result by 2 to obtain the total distance that Tom ran: (1,142.857142)/2 = 571.428571 yards.

Since the pig ran at 3/4 of the speed of Tom, the total distance traveled by the pig is 428.571428 yards.

How did Sam Loyd derive his simple rule? We cannot be certain, but apparently, he may have solved the problem the hard way. In other words, he delved into the intricacies of the calculus to find the shortcut that he gave to solve the problem. Loyd published the shortcut solution in the *Cyclopedia*. Loyd's discovery of that simple rule illustrates just how good a mathematician he was.

In his book *Ingenious Mathematical Problems and Methods* (problem 74), L. A. Graham shows how the problem can be solved by calculus. His version of the puzzle involves a dog chasing a cat. The cat is 60 yards north of the dog. The dog runs 5/4 times as fast as the cat. The cat runs due east. At all times, the dog is running directly toward the cat. Graham's calculations eventually lead to another simple rule that differs from Loyd's to determine the distance traveled by the pursued.

Graham's rule is

$$\frac{\left(\frac{5}{4}\right)60}{\left(\frac{25}{16}\right)-1}$$

In words, the rule found by Graham is as follows: the distance traveled by the pursued equals the initial distance between pursuer and the pursued multiplied by the ratio of their speeds divided by a number one less than the square of that ratio.

Graham's rule tells us that the cat traveled 133.333333 yards. Since the dog is traveling 5/4 times faster than the cat, the dog travels 166.666666 yards before he catches the cat.

In Loyd's version of the puzzle, Graham's rule is

$$\frac{\left(\frac{4}{3}\right)250}{\left(\frac{16}{9}\right)-1}$$

The rule gives 428.571428 yards as the distance traveled by the pig. As Tom is traveling 4/3 times faster than the pig, Tom travels 571.428571 yards.

Perhaps Loyd could have pointed out that an even simpler rule applied that can be used to determine the total distance traveled by Tom. This simplified rule, which to the best of my knowledge has not been previously published, is this: consider *only* the situation whereby Tom and the pig run in a straight line in the same direction before Tom captured the pig. If that were the case, Tom would catch the pig after running 1,000 yards. The ratio of both their speeds is 4 to 3. Now obtain 4/7 of 1,000 yards. The answer is 571.428571 yards. That gives the total distance traveled by Tom when chasing the pig as it ran directly toward the left-hand bottom corner of the field. The total distance traveled by the pig is 3/7 of 1,000, or 428.571428 yards. That equals the total distance traveled by the pig as he is being chased by Tom.

Applying this new rule to Graham's version of the problem, we find that the dog would catch the cat after 300 yards if both dog and cat were running in the same direction. The ratio of their speeds is 5 to 4. Obtain 5/9 of 300. The result is 166.666666 yards. That is the total distance traveled by the dog before he catches the cat. The total distance traveled by the cat as he is being chased by the dog is 4/9 times 300, or 133.333333 yards.

In his book *Puzzles and Curious Problems* (problem 210), Henry Ernest Dudeney gives his version of the problem. The pig is chased by a man 100 yards south of the pig. The man is traveling twice as fast as the pig. The pig runs due west. At all times, the man is running directly toward the pig. How far does each travel before the pig is caught?

Dudeney gives the correct answer to the problem but does not explain how to find that solution. However, in the preface to the book, Dudeney's wife, Alice, gives the same formula for solving the puzzle as Graham does in his book. In Dudeney's version, the man travels a total distance of 133.333333 yards, and the pig runs a total of 66.666666 yards.

I first came across Loyd's puzzle of Tom the Piper's Son when I was a teenager. I was fascinated by Loyd's simple rule that can be used to solve the problem.

There are other formulas that can used to solve the Tom the Piper's Son puzzle. Here is a simple one that I discovered. Let A/B equal the ratio of the speed between the man and the pig. In Loyd's version of the puzzle, A/B would equal 4/3. Let C equal the initial distance between the man and the pig before the chase begins. In Loyd's version of the puzzle, C equals 250 yards. The total distance traveled by the man equals $(A^2C)/(A^2 - B^2)$. In Loyd's version of the puzzle, this equals $(4 \times 4 \times 250)/(4^2 - 3^2)$, or 571 and 3/7 yards.

The total distance traveled by the pig is $(ABC)/(A^2 - B^2)$. This equals $(4 \times 3 \times 250)/(4^2 - 3^2)$. This equals 428 and 4/7 yards.

The same formulas can, of course, be used to solve Dudeney's version or Graham's version (or indeed any version of the puzzle).

Finally, I give two problems that are usually solved by calculus, but they can be solved also by non-calculus means.

The first problem is as follows. A farmer has 100 feet of fencing and wishes to build a rectangular enclosure, using the barn wall as one of the sides of the enclosure. What should the dimensions of the enclosure be so that the area is maximized using the 100 feet of fencing at his disposal?

The second problem is as follows. Suppose that the regulations for posting a rectangular box state that the sum of the length and the girth cannot exceed 108 inches. What is the largest rectangular box in volume one can send by mail?

Solutions

1. If the farmer knew a little bit of arithmetic, he could solve this problem almost instantly without calculus.

    First, let us assume he is aware that a square with a side equal to x is equal in area to x^2. A rectangle that has sides equal to $x + 1$ and $x - 1$ (thereby preserving the length of the perimeter) has an area of $x^2 - 1$. A rectangle that has sides equal to $x + 2$ and $x - 2$ has an area of $x^2 - 4$, and so on. It will soon become apparent to the farmer that of all the possible rectangles with a given area, it is the square that has the largest area.

    Thus, he should imagine a rectangular fence built up against the wall of the barn so that the rectangle enclosed by fencing would resemble half of a square. One side of the rectangular fence would have a length of 50 feet. Each of the other two sides would have a length of 25 feet. Now he imagines that on the other side of the wall, there is a similar rectangular area. In other words, he imagines a square measuring 50 feet by 50 feet, with the straight barn wall dividing the square into two equal rectangles, each measuring 50 feet by 25 feet (see the figure below). Since the square measuring 50 feet by 50 feet maximizes the area with 200 feet of fencing, half of that maximum area equals the rectangle measuring 50 feet by 25 feet, with 100 feet of fencing available.

Thus, this rectangle of 50 feet by 25 feet, with an area of 1,250 square feet, maximizes the area with 100 feet of fencing available:

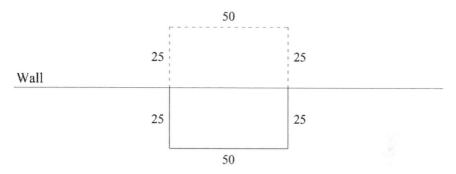

The problem can, of course, be tackled by calculus as follows: Let the total length of fencing equal l. Let x equal the length of each of the two sides of the rectangle that are perpendicular to the barn wall. Thus, the side of fencing that is parallel to the barn wall can be expressed as $l - 2x$.

We can then write the following:

$$Area\ of\ rectangle = A = x(l - 2x) = xl - 2x^2$$

We wish to make the enclosed area a maximum, so to find the value of x, we differentiate A with respect to x and equate the result to 0.

Thus, we obtain $l - 4x = 0$, or $l = 4x$, or $1/4\ l = x$.

Thus, $l - 2x = 1/2\ l$.

In our problem, $l = 100$. Thus, $1/2\ l = 50$.

This tells us that the maximum area of the rectangle is achieved when the side of fencing that is parallel to the barn wall is 50 feet in length. Each of the two sides that are perpendicular to the barn wall will then be 1/4 of l, or 25 feet in length. The area of the enclosure then is 50×25, or 1,250 square feet.

Calculus can also be used to find the solution to the latter problem, that is, the problem of finding the sides of a rectangle that gives the maximum area.

Suppose the perimeter of the rectangle is P. Suppose also that one of the sides of the rectangle equals x. The other side length is $(P - 2x)/2$. This allows us to write

$$Area\ of\ rectangle = A = x\left(\frac{P - 2x}{2}\right) = \frac{1}{2}Px - x^2$$

The area of the rectangle is maximized by differentiating A with respect to x and making the result equal to zero. Thus, we obtain $1/2P - 2x = 0$, which makes

$$\frac{1}{4}P = x$$

In other words, since there are 4 sides to a rectangle, the area of the rectangles is at a maximum when each side is equal to 1/4 P, which means when the rectangle is a square.

2. The second problem states that the sum of the length and the girth of a rectangular box sent through the postal system cannot exceed 108 inches. What is the largest rectangular box in volume that one can send by mail?

To obtain the volume of the largest rectangular box, we use the result we obtained in the previous problem: The area of a rectangle is at a maximum when each side is equal to a square. Therefore, it makes sense that the area of the end of the rectangular box should be in the form of a square, say, x^2. If the length of the box is l, the volume of the box equals lx^2.

We know that the volume of the largest box under the postal regulations means that $4x + l = 108$, or $l = 108 - 4x$. Substituting this into our equation for the maximum volume of the box gives maximum volume $= V = x^2(108 - 4x)$ $V = 108\,x^2 - 4x - 4x^3$.

We then differentiate V with respect to x and obtain $V = 216x - 12x^2$.

We now set $V^1 = 0$.

Thus, $0 = 216x - 12x^2$.

This equation can be reduced to $0 = 18x - x^2$.

Rearranging and dividing by x gives $x(x - 18) = 0$.

Therefore, x must equal 0 or x equals 18.

We see that x cannot equal 0 because it does not fit the problem's conditions. Therefore, x must equal 18.

We substitute this value of x in our equation for the maximum volume of the box:

$$l = 108 - 4x$$
$$l = 108 - 72$$
$$l = 36$$

Therefore, the maximum volume of a rectangular box that can be sent through the postal system, where the length plus the girth does not exceed 108 inches, is

End sides of the rectangular box = 18 inches by 18 inches

Length of box = 36 inches

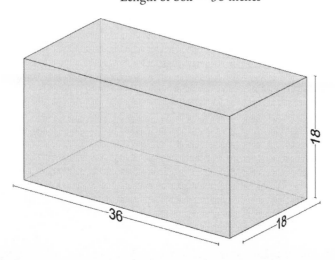

This problem (and similar problems) can be solved without resorting to calculus by using the following simple rule: simply divide 108 by 6. (One always divides the maximum sum of the length and girth of the rectangular box by 6.) The answer, 18, is the width of the square end. Then multiply 18 by 2, obtaining 36. Therefore, 36 is the length of the rectangular box.

That's all there is to it!

Therefore, the dimensions of the rectangular box that gives the maximum volume is $18 \times 18 \times 36 = 11,664$ cubic inches.

This simple rule applies no matter what the dimensions of the box are.

The power of the calculus or non-calculus solution can be seen in just a few lines above. If we did not know calculus or did not know the non-calculus solution, we could play around with numbers and see if we can get the dimensions of the box that give a maximum volume. For example, if we made each of the side ends of the box 18.5 inches, the length of the box would be $108 - 4 \times 18.5 = 34$ inches. The volume of the box would then be $18.5 \times 18.5 \times 34 = 11,636.5$ cubic inches. That is 27.5 cubic inches less than the maximum volume. If each of the end sides of the rectangular box was 18.75 inches, the length of the box would be 33 inches. The volume then would equal $18.75 \times 18.75 \times 33 = 11,601.5625$ cubic inches. That is 62.4375 cubic inches short of the maximum volume. If each of the end sides of the rectangular box was 17.9 inches, the length of the box would be 36.4 inches. The volume then would equal $17.9 \times 17.9 \times 36.4 = 11,662.924$ cubic inches. That is just 1.076 cubic inches short of the maximum box with the given constraints. We could continue in this vein, hoping that we would find the maximum volume of the rectangular box. However, the calculus solution, or the non-calculus solution given here, can give us the maximum volume almost instantly, demonstrating their superiority over the trial-and-error method.

CHAPTER 12

Two Lightning Calculation Tricks and Sundry Other Matters

For centuries, lightning mental calculators have attracted crowds and admirers. These mental calculators demonstrated extraordinary abilities in their shows. Sometimes these individuals apparently did not know how they did what they did. Volumes have been written attempting to explain their incredible feats.

I will give here one of their more spectacular feats, namely, the calendar trick, that is, naming the day of the week for any date called out.

I will then discuss a method of calculating the date of the *Paschal full moon* in a given year, which is used in the method to decide the date of Easter for that year. I will show how to mentally calculate the date of Easter for any year by combining the calendar trick with the method of calculating the Paschal full moon. With practice, the trick can be done surprisingly quickly and will undoubtedly impress an academic audience, which is aware of just how difficult it is to determine the date of Easter by traditional methods.

The Calendar Trick

First, I will give the key numbers for each of the 12 months of the year. The key numbers for each month appear below each month:

Jan	Feb	Mar		April	May	June		July	Aug	Sep		Oct	Nov	Dec
1	4	4		0	2	5		0	3	6		1	4	6

(I find it is best to remember these key numbers as follows. The key numbers for the first three months equal the square of 12. The key numbers for the next three months equal the square of 5. The key numbers for the next three months equal the square of 6. Finally, the key numbers for the last three months equal the square of 12 plus 2.)

Step 1: We first calculate the day of the week for any date in the 1900s. For other centuries, only one final adjustment is needed at the end of the procedure.

Consider the last two digits of the year as a single number. Divide it mentally by 4 and keep only the quotient in mind. You now add the quotient to the year number. For example, suppose the year is 1942. Your calculations are as follows. Four goes 10 times into 42. Retain the number 10 in your head and ignore the remainder. Add 10 to 42, obtaining 52. Cast out 7s. In other words, take as many 7s as possible from 52 and remember only the remainder. One obtains 3. (This procedure of casting out 7s is usually called modulo 7.) Therefore, in this example, we remember only 3.

Step 2: To the result of the preceding step, add the month's key number. If possible, cast out 7s.

Step 3: To the preceding result, add the day of the month. Cast out 7s if possible. The resulting digit gives the day of the week, counting Saturday as 0, Sunday as 1, Monday as 2, and so on, to Friday as 6.

Step 4: If the year is a leap year and the month is January or February, go back one day from the final result.

The first step automatically alerts you to leap years. Leap years are multiples of 4, and any number is a multiple of 4 if its last two digits are. Therefore, if there is no remainder when you divide by 4, you know it is a leap year. (Keep in mind that in the Gregorian calendar, 1800 and 1900, although multiples of 4, are not leap years, while 2000 is. The reason is that the Gregorian calendar provides that a year ending in two zeros is a leap year *only* if it is evenly divisible by 400.)

The procedure just explained is used for dates in the 1900s. Only very trivial final adjustments need be made for dates in other centuries. For the 1800s, go forward two days in the week from the final result. For dates in the 2000s, go back one day from the final result. For dates in the 1700s, from 1753 on, go forward four days from the final result.

It is best not to allow dates in the 1700s prior to and including 1752 because of the confusion caused by the shift in the calendar that happened in that year. On September 14, 1752, the old Julian calendar was dropped in most of Europe, and the new Gregorian calendar was adopted in its place. Pope Gregory, acting on the advice of mathematicians, authorized the dropping of 10 days from the calendar so that the new Gregorian calendar would be as correct as possible. The change was implemented so that Thursday, October 4, 1582, of the Julian calendar was followed by Friday, October 15, 1582, of the Gregorian calendar.

This change was done in most of Europe in 1582, but in the English-speaking world, the change was not made until 1752. By that year, the error had accumulated to 11 days (because 1700 in the old Julian calendar was a leap year, while in the Gregorian calendar, it is not). In 1752, when the calendar was changed in the English-speaking world, the day after September 2 was called September 14. If you are going to do the calendar trick for dates in the 1700s, do it for dates beginning with 1753 on. For dates in the 1700s, from 1753 on, perform the four-step procedure given here and then go forward four days in the week from the final result.

A few examples will make the procedure clear.

Suppose someone asks you what day of the week November 22, 1963, fell on. (That was the day President John F. Kennedy was assassinated.) Your mental calculations are as follows. Divide 63 by 4 (obtaining 15) and ignore the remainder. Add 15 to 63, obtaining 78. Cast out 7s. You are left with 1. The month we are dealing with is November. The key number for November is 4. Add the preceding 1 to 4, obtaining 5.

Casting out 7s still leaves 5. The day of the month we are dealing with is 22. Add 5 to 22, obtaining 27. Casting out 7s reduces this to 6. The 6 corresponds to Friday. Therefore, you know that November 22, 1963, fell on a Friday.

Suppose you were asked to calculate the day of the week for November 22, 1863. You perform the calculations exactly as described above for November 22, 1963. When you reach the final calculation, which is 6, go forward two days so that you obtain 8. Casting out 7s reduces this to 1. Therefore, you know that November 22, 1863, fell on a Sunday.

If you were asked what day of the week November 22, 2063 falls on, you perform the calculation as described above for November 22, 1963. Then go back one day from your last answer. Thus, you know that November 22, 2063, falls on a Thursday.

Similarly, if you calculate the day of the week for November 22, 1763, you perform the calculation as you do for November 22, 1963. You then go forward four days from your final result, 6, obtaining 10. Casting out 7s reduces this to 3. You know that 3 corresponds to Tuesday. Therefore, November 22, 1763, fell on a Tuesday.

Suppose you are asked what day of the week July 4, 1776, was. (That was the day of the signing of America's Declaration of Independence.)

Your mental calculations are as follows. Divide 76 by 4, obtaining 19. (The fact that zero is the remainder after you divide by 4 alerts you to the fact that 1776 was a leap year, but you need not make any adjustment because the month you are dealing with is *not* January or February.) Now add 76 and 19, obtaining 95. Casting out 7s leaves you with 4. The day of the month we are dealing with is July. The key number for July is 0. Add the preceding 4 to 0, obtaining 4. Casting out 7s still leaves 4. The day of the month we are dealing with is 4. Add 4 to 4, obtaining 8. Casting out 7s reduces this to 1. Now, because you are calculating a date in the 1700s, from 1753 on, you go forward 4 days from your final result. Thus, you add 4 to 1, obtaining 5. The 5 corresponds to Thursday. Therefore, you know that July 4, 1776, fell on a Thursday.

No adjustments for leap years are required in the years used in these examples, because only one year, 1776, was a leap year, but the month we were dealing with was *not* January or February.

Let us now suppose you are asked what day of the week January 10, 1936, fell on.

Your mental calculations are as follows. Divide 36 by 4, obtaining 9. You now know (because 36 is evenly divisible by 4) that you are dealing with a leap year. No adjustment need be made if the month is *not* January or February. But we are dealing with January. Now add 9 to 36, obtaining 45. Casting out 7s gives 3. The month we are dealing with is January. The key number for January is 1. Add the preceding 3 to 1, obtaining 4. Casting out 7s still leaves 4. The day of the month we are dealing with is 10. Add 4 to 10, obtaining 14. Casting out 7s reduces this to 0. The number 0 corresponds with Saturday. Because you are calculating a date in January in a leap year, go back one day from your result. This gives Friday. Therefore, you know that January 10, 1936, fell on a Friday. You also know, without doing almost any calculating, that January 10, 1836, fell on a Sunday. You also know that January 10, 2036, will fall on a Thursday and that January 10, 2136, will fall on a Tuesday.

The Gregorian calendar repeats every 400 years. (This is because there are $400 \times 365 + 97$, or 146,097 days, in a 400-year period. The number 146,097 is evenly divisible by 7.)

Therefore, dates in the 1700s (from 1753 on) will fall on the same day of the week for similar dates in the 2100s. Dates in the 1800s will fall on the same day of the week for similar dates in the 2200s. Dates in the 1900s will fall on the same day of the week for similar dates in the 2300s. Dates in the 2000s will fall on the same day of the week for similar dates in the 2400s, and so on. Thus, even if someone gives you a date, say, in the year 2400, you know that that is 400 years ahead of the year 2000. In that case, you do the calculation as normal for dates in the 2000s. No adjustment need be made at the end because similar dates fall on the same day of the week in each of the two centuries, 2000 and 2400.

This trick may appear complicated, but with practice, the interested reader will easily master it.

The Date of Easter

You can then use the calendar trick to calculate the date of the Paschal full moon, whose date decides the date of Easter. Unbelievably, once you have mastered this trick, the calculation of the date of Easter for any year in the 1900s and 2000s is remarkably easy.

The date of Easter falls on the first Sunday after the first Paschal full moon, which falls on or after the vernal or spring equinox, and, for the purpose of calculating the date of Easter, is assumed to fall on March 21 each year. The Paschal full moon is an abstract entity that is given a designated date that is obtained as the result of a mathematical equation.

For calculating the date of Easter, the Paschal full moon usually (but not always) coincides with the *astronomical full moon*. (The astronomical full moon is, of course, based solely on *astronomical* calculations.)

Interestingly, these rules state that the vernal equinox is fixed on March 21, even though from the years 2008 through 2101, at European longitudes, the spring equinox will occur no later than March 20.[1]

Because of the rules, it is easily proved that the earliest possible date for Easter is March 22 and that the latest possible date is April 25.

There have been several formulas devised to calculate the date of Easter. These formulas, while relatively simple in nature, are tedious and laborious to apply and are not suitable for calculating the date of Easter mentally.

I find the formula devised by the English mathematician John Conway (1937–2020) to be the best suited for the mental calculation of Easter. Conway devised the following relatively simple rule for calculating the date of the Paschal full moon. Once the date of the Paschal full moon is calculated, it is then a simple matter to find the date of Easter in that year.

Conway's formula may initially appear difficult, but it is surprisingly simple to use. It can be used to figure out the date of the Paschal full moon in both Gregorian and Julian calendars.

Here is Conway's formula:

$$(\text{April } 19 = \text{March } 50) - (11\,G + C) \text{ modulo } 30$$

However, when the formula gives April 19, you should take April 18 instead, and when the formula gives April 18 and G either equals or is greater than 12, you should take April 17. In the formula, G stands for the *golden number*, which equals the $\text{Year}_{\text{modulo 19}} + 1$. (Always remember to add the 1.) C stands for the *century term*, which is +3 for all Julian years:

−4 for the 1500s and 1600s
−5 for the 1700s and 1800s
−6 for the 1900s, 2000s, and 2100s

The formula for finding the value of C in a Gregorian year Hxx is

$$C = -H + \lfloor H/4 \rfloor + \lfloor 8(H + 11)/25 \rfloor$$

Thus, the value of C in the 2000s is

$$C = -20 + \lfloor 20/4 \rfloor + \lfloor 8(20 + 11)/25 \rfloor,$$

which gives

$$C = -20 + 5 + 9$$

Thus,

$$C = -6.$$

We will now use Conway's formula to calculate the Paschal full moon for the year 2022. Conway's formula is

$$(\text{April } 19 = \text{March } 50) - (11\,G + C) \text{ modulo } 30,$$

$$\text{which equals } (\text{April } 19 = \text{March } 50) - (99 - 6) \text{ modulo } 30,$$

$$\text{which equals } (\text{April } 19 = \text{March } 50) - (3) \text{ modulo } 30,$$

$$\text{which equals April } 16.$$

Thus, in 2022, the Paschal full moon falls on April 16.

We next use the calendar trick to calculate the day of the week April 16 falls on in 2022. We find it is a Saturday. Therefore, the next day, Sunday, April 17, is the date of Easter.

If we wish to figure out the Paschal full moon for the year 1818, we find that

$$(\text{April } 19 = \text{March } 50) - (11 \times 14 + C) \text{ modulo } 30,$$

$$\text{which equals } (\text{April } 19 = \text{March } 50) - (154 - 5) \text{ modulo } 30,$$

$$\text{which equals } (\text{April } 19 = \text{March } 50) - (149) \text{ modulo } 30,$$

$$\text{which equals } (\text{April } 19 = \text{March } 50) - (29) \text{ modulo } 30,$$

$$\text{which equals March } 21.$$

We then use the calendar trick to figure out that the Paschal full moon in 1818 fell on a Saturday. The following day, March 22, 1818, was Easter.

We now use Conway's formula to calculate the Paschal full moon for the year 2038:

$$(\text{April } 19 = \text{March } 50) - (11 \times 6 + C) \text{ modulo } 30,$$

$$\text{which equals } (\text{April } 19 = \text{March } 50) - (66 - 6) \text{ modulo } 30,$$

$$\text{which equals } (\text{April } 19 = \text{March } 50) - (60) \text{ modulo } 30,$$

$$\text{which equals } (\text{April } 19 = \text{March } 50) - (0) \text{ modulo } 30,$$

$$\text{which equals April } 19.$$

We know that when the formula gives April 19, we take April 18 instead. Therefore, in 2038, the Paschal full moon will fall on April 18. We use the calendar trick to figure out that this is a Sunday. Therefore, the *following* Sunday, April 25, is Easter. This is the latest possible date for Easter.

For the quick mental calculation of Easter, it is probably best to restrict dates to the Gregorian calendar in the 1800s, 1900s, 2000s, and 2100s.

However, to those who aspire to doing greater things, the mental calculation of Easter can be performed, without too much added effort, for dates in the Julian calendar. The following are a few examples.

Suppose we wish to use Conway's formula to calculate the Paschal full moon for the year 1542. (That year, the Julian calendar was in operation.) We first use Conway's formula to determine the Paschal full moon for 1542. Since the date is in the Julian calendar, C is equal to +3:

$$(\text{April } 19 = \text{March } 50) - (11 \times 4 + 3) \text{ modulo } 30,$$

$$\text{which equals } (\text{April } 19 = \text{March } 50) - (44 + 3) \text{ modulo } 30,$$

$$\text{which equals } (\text{April } 19 = \text{March } 50) - (47) \text{ modulo } 30,$$

$$\text{which equals } (\text{April } 19 = \text{March } 50) - (17) \text{ modulo } 30,$$

$$\text{which equals April } 2.$$

Thus, the date of the Paschal full moon in the Julian calendar in the year 1542 was April 2. How do we determine what day of the week April 2 fell on?

Here is how we do it.

In the 1500s, the Julian calendar was 10 days out of sync with the Gregorian calendar. Thus, we add 10 to April 2, obtaining April 12. Using the calendar trick, we find that April 12, 1542, was a Sunday under the proleptic Gregorian calendar. Therefore, April 2 in the Julian calendar was also a Sunday. Recall that that was the date of the Paschal full moon. Therefore, the following Sunday, April 9, was Easter in 1542.

How do we know how many days to add to the Julian date to convert it to the equivalent Gregorian date? Use the following simple formula. Let A be the first two digits of the year. Then calculate A × 0.75 − 1.25. Disregard any digits in the answer to the right of the decimal point. For example, to find how many days to add in the 1500s, our calculations are 15 × 0.75 − 1.25 = 10. For the 1400s, the calculations are 14 × 0.75 − 1.25 = 9.25. Disregard the decimal digits in 9.25 to obtain 9 as the final answer. Thus, one adds 9 days for dates in the 1400s, and so on. The following is a handy table for reference that shows how many days out of sync the Julian calendar is compared to the Gregorian calendar:

1500s	1600s	1700s	1800s	1900s	2000s	2100s
10 days	10 days	11 days	12 days	13 days	13 days	14 days

Let us now determine the date of the Paschal full moon and the date of Easter for the year 999 CE. We are dealing here with the Julian calendar.

Therefore, we have

$$\text{(April 19 = March 50)} - (11 \times 12 + 3) \text{ modulo } 30,$$

$$\text{which equals (April 19 = March 50)} - (132 + 3) \text{ modulo } 30,$$

$$\text{which equals (April 19 = March 50)} - (15) \text{ modulo } 30,$$

$$\text{which equals (April 19 = March 50)} - (15) \text{ modulo } 30,$$

$$\text{which equals April 4.}$$

Thus, the Paschal full moon in 999 was on April 4 in the Julian calendar. Our table above does not include the 900s. But it is easy to use our formula to calculate how many days to add to 4 to convert the Julian date to the proleptic Gregorian date. Thus, $9 \times 0.75 - 1.25 = 5.5$. Drop all digits to the right of the decimal point. This leaves us with 5. Thus, we add 5 days to April 4 in the Julian calendar, obtaining April 9 in the proleptic Gregorian calendar. Thus, April 4 (Julian calendar) fell on the same day of the week as April 9 (Gregorian calendar).

Using the calendar trick, we calculate that April 9 in 999 in the proleptic Gregorian calendar was on a Tuesday. Therefore, April 4 (Julian calendar) was on a Tuesday. The following Sunday, April 9, was Easter.

We can now reveal a few statistics concerning the date of Easter in the Gregorian calendar.[2]

If the Paschal full moon falls on March 21 and if that day is a Saturday, the next day, March 22, is Easter Sunday. This is the earliest date for Easter and is also the rarest Easter date. Easter falls on March 22 about once every 207 years on average. Easter fell on this date in 1598, 1693, 1761, and 1818. Then there is a gap of 467 years before Easter will next fall on March 22, in 2285. Thereafter, Easter will fall on March 22 in 2353, 2437, 2505, and then there will be another gap of 467 years before it happens again, in 2972.

The third-rarest date for Easter is March 23. It falls on this date about once every 105 years on average. Easter fell on March 23 in 1704, 1788, 1845, 1856, 1913, and 2008. It will next occur in 2160 and then again in 2228.

When Easter falls on March 24, the Sunday preceding Easter (Palm Sunday) falls on St. Patrick's Day, which is celebrated each year on March 17. This happened in 1799 and 1940. Then there is a gap of 451 years before it happens again, in 2391. Thereafter, it will occur in 2475, 2543, and 2695. Easter falls on March 24 once every 70 years on average.

The latest possible date for Easter is April 25. This is also the second-rarest date for Easter, which falls on this date about once every 136 years on average. The three last occasions Easter fell on April 25 were 1734, 1886, and 1943. Easter will not fall on April 25 again until 2038. Thereafter, Easter will be on April 25 in 2190, 2258, and 2326.

The most frequent date for Easter is April 19, with April 18 running a close second. Easter falls on April 19 on average about once every 25 years. It falls on April 18 on average about once every 28 years.

The cycle of Easter dates in the Gregorian calendar is 5,700,000 years.

The probability that any random date of Easter falls in March is 53,181/228,000, which equals 23.325 percent.

The following are a few statistics concerning Easter in the Julian calendar.

The cycle of Easter dates in the Julian calendar is 532 years, of which the earliest is March 22. It falls on this date on average about once every 133 years. This happened in 1668, 1915, and again in 2010. It will not fall on March 22 again until 2105. The latest date for Easter in the Julian calendar is April 25. This occurred in 1641, 1736, and 1983. It will not happen again until 2078. March 22 and April 25 are the two rarest dates for Easter in the Julian calendar, occurring on these dates about once every 133 years on average.

The most frequent dates for Easter in the Julian calendar are, with equal probability, March 28 and 31 and April 3, 5, 6, 8, 11, 14, 16, and 19. The exact probability in all of these cases is 34/133, or about once every 26 years.

Because of the 532-year cycle in the Julian calendar, once one has established a particular date for Easter, then 532 years later, the same date for Easter will apply. For example, Easter fell on March 22 in the Julian calendar for the years 414, 946, 1478, and 2010. It will also fall on March 22 in 2542 and 3074 and so on.

In 2025, Easter in the Julian calendar will be on April 7, and in the Gregorian calendar, Easter will be on April 20. Both Easter dates fall on the same day. That will happen again in 2028 and 2031.

Of course, calculating Easter dates into the far future is merely an academic exercise. It provides endless fun and fascination for those interested in numbers and calculations. However, it has no practical relevance. In practical terms, the slowing down or speeding up of Earth's rotation affects the length of a "day." That in turn affects the length of a "year" and makes redundant any serious attempt to calculate the date of Easter into the far-distant future.

CHAPTER 13

Parallels in the Lives of Sam Loyd and Martin Gardner

The following is an original list of parallels published here for the first time concerning the American puzzle genius Sam Loyd and the great popularizer of recreational mathematics Martin Gardner. These curiosities were recently passed on to me by e-mail by the remarkable numerologist Dr. Moogle.

Let me first point out that Sam Loyd was born on January 30, 1841, and died on April 10, 1911. Martin Gardner was born on October 21, 1914, and died on May 22, 2010.

Now to the parallels between the two men.

Sam Loyd was famous for his enormous interest in chess and recreational mathematics.
Martin Gardner was famous for his enormous interest in chess and recreational mathematics.
In the nineteenth century, Sam Loyd made famous many puzzles in recreational mathematics.
In the twentieth century, Martin Gardner made famous many puzzles in recreational mathematics.
Sam Loyd was born in the 41st year of the nineteenth century.
Martin Gardner was born in the 14th year of the twentieth century.
Sam Loyd was born in the $(4^2 + 5^2)$ year of the nineteenth century.
Martin Gardner was born in the $(1^2 + 2^2 + 3^2)$ year of the twentieth century
Sam Loyd died on the $(45 + 55)$ day of the year.
Martin Gardner died on the $(66 + 76)$ day of the year.
Sam Loyd was born on a Saturday.
Martin Gardner died on a Saturday.
Sam Loyd was born in Philadelphia and later moved to New York City.
Martin Gardner was born in Oklahoma and later moved to New York City.
Sam Loyd was born on the 28 + 2 day of the year in 1841. There are 282 primes less than 1841.
Martin Gardner was born on the 293 + 1 day in the year 1914. There are 293×1 primes less than 1914.

The initials of Sam Loyd are the 19th and 12th letters in the alphabet. Those two numbers sum to 31. Subtract 1 from 31 to obtain 30. Sam Loyd was born on the 30th day of the month.

The initials of Martin Gardner are the 13th and 7th letters in the alphabet. Those two numbers sum to 20. Add 1 to 20 to obtain 21. Martin Gardner was born on the 21st day of the month.

In 1853, Sam Loyd was 12 years old. In 1866, he was 25 years old, and in 1893, he was 52 years old. The three numbers 12, 25, and 52 sum to 89.

In 1935, Martin Gardner was 21 years old. In 1966, he was 52 years old, and in 1939, he was 25 years old. The three numbers 21, 52, and 25 sum to 98.

Sam Loyd wrote a column for Scientific American *magazine.*

Martin Gardner wrote a column for Scientific American *magazine.*

Sam Loyd sprung a major hoax concerning the origin of the game of tangrams.

Martin Gardner wrote a major hoax column in Scientific American.

Sam Loyd had a brother named Thomas.

Martin Gardner had a son named Tom.

In 1856, Loyd achieved his breakthrough with his chess puzzles. That year, the New York Clipper *ran a competition for the best chess problem given to its paper. Sam Loyd won the competition, which led to his hugely successful career as a chess puzzlist.*

In 1956, Martin Gardner made his breakthrough in recreational mathematics. That year, Scientific American *published an article by Martin Gardner that led to his hugely successful monthly column on mathematical recreations in that same magazine.*

CHAPTER 14

Dr. Moogle Gives a Numerological Analysis of the 1966 FIFA World Cup Tournament in England

On the first day of December 2020, I received an e-mail from Dr. Moogle's daughter, Anna. The e-mail said that Dr. Moogle would be in London, in the United Kingdom, on December 11. She mentioned the hotel that they would be staying in. Anna went on to say that if I wished, I could arrange a private interview with her father in his hotel room. She said her father was keen to give me a numerological analysis of the 1966 FIFA World Cup Tournament, which was held in England, saying that the information would be unique and original. Anna believed that my readers would be most interested in what her dad had to say.

I booked the interview and left by airplane for London on Friday, December 11. I booked into a hotel in the north of London. The next day, I took a taxi to the hotel, where Dr. Moogle was staying, in the West End of England's capital.

The following is a summary of an exceptionally long and interesting interview I had with the great man himself.

O.O'S: May I ask what interesting details of the FIFA 1966 World Cup you wish to convey?

Dr. Moogle: Well, please let me begin by informing your readers of a few facts to help them appreciate the astonishing information I will be revealing to you in this interview.

First, may I point out that England won the World Cup Final in Wembley Stadium, London, on Saturday, July 30, 1966. England's opponents were West Germany. England won by four goals to two, after extra time was played. The game was two goals each at full time. Geoff Hurst, who was born on Monday, December 8, 1941, scored three goals (known to soccer fans as a hat trick) in the Final. He is the only soccer player to date to score three goals in a World Cup Final. The England team that won the World Cup Final was captained by the late Bobby Moore.

O.O'S: I am aware of that. Geoff Hurst is world-famous because of his hat trick in the Final. I should tell you that I watched the 1966 World Cup Final on television with my twin brother, Michael. It is one of the most famous World Cup Finals of all time. Are you going to tell me that you have some interesting info on the 1966 World Cup Tournament?

Dr. Moogle: Let's begin at the beginning. First, before we start talking about soccer, please let me point out some more curiosities that have an English flavor. Your readers may like them. The patron saint of England is *George*. The symbol of England is the *Rose*. Using the usual alphabet code, where A equals 1, B equals 2, and so on, the words ENGLAND, GEORGE, and ROSE each sum to 57. Incidentally, using the same code, the letters of England's capital city, LONDON, sum to 74, which equals $5^2 + 7^2$.

O.O'S: This is interesting. I must admit I never heard of that before. I am sure many English people will find that interesting also. Have you other curious facts?

Dr. Moogle: Yes, I do. Quite a number. England's association with the number 57 becomes obvious when one recognizes that the first letter of the word ENGLAND is the 5th letter of the alphabet and that the word ENGLAND has 7 letters. Because of England's association with the number 57, one would expect that that number would have some significance in English history. Well, it so happens that it does! The military surrender of Nazi Germany (or to give it its more formal term, the *German Instrument of Surrender*) occurred on May 7, 1945. That was the legal instrument to bring Germany's involvement in the war in Europe to an end. That date may be written as 5/7. Of course, it may be said that on that day, England won the war.

O.O'S: That is most interesting.

Dr. Moogle: I am pleased you like it. There is much more interesting information to come. I have some remarkably interesting information on the 1966 World Cup Final. I was born in India, then moved to New York for about ten years, before moving to France in my early teens. I have always liked Soccer. Most of my friends growing up in the *Big Apple* liked baseball or basketball. However, I was always my own man, and I happened to like soccer. I played a little when I was younger. As you know, I am very modest, but I must admit that I was a great soccer player! I loved the *beautiful game*. Of course, I was always on the lookout for some curiosities involving the statistics of the game that others missed. I have in my files a lot of original curiosities that I discovered concerning the World Cup in 1966. I will give you these curiosities if you wish, as I think your readers might like this info. I know that your readers in England will particularly enjoy the curiosities I give.

O.O'S: Okay. Lay it on me. I might be able to use some of it in my forthcoming book on mathematical diversions.

Dr. Moogle: Saturday, July 30, 1966, is a date etched on the minds of many English soccer fans. That was the glorious summer day long ago that England became World Soccer Champions for the first time (and only time to date) at the game they gave to the world.

O.O'S: That much, I do know.

Dr. Moogle: When I have completely given you my numerological analysis of the 1966 World Cup, you might agree that 1966 was a year in which England was destined for glory in that competition. I will explain. First, we will use the usual alphabet code where A equals 1, B equals 2, C equals 3, and so on. Now consider the following: the World Cup Final in 1966 was played at Wembley Stadium in London, UK. This stadium was previously known as *Empire* Stadium. The game was played in the city in which the river *Thames* flows. The captain of the England team that day was the late Bobby *Moore*. Using the code above, one finds that the sum of the letters in each of the three words, *Empire*, *Thames*, and *Moore*, equals 66.

O.O'S: That is new to me. It almost seems that something significant was going to happen in the year '66.

Dr. Moogle: Exactly! The year 1966 was a glorious year for England. The England captain that July day in 1966 was Bobby Moore (1941–1993). Consider this. Print the name BOB MOORE. Using the earlier alphabet code, substitute the relevant numbers in place of each letter. Sum the values of the letters in each of the two names. The answer is surprising.

O.O'S: Let me do it. [I jotted down the name Bob Moore, and underneath it, I substituted the proper numbers for each of the letters. I then added the sum of the letters in each of the two names.] This is what I obtained:

BOB MOORE
19 66

O.O'S: That's wonderful! Bobby Moore is a legendary figure in England. He died in 1993. I never saw that curiosity before. I do not think anyone ever spotted it. I am sure Bobby Moore would have enjoyed that curiosity if he had known of it.

Dr. Moogle: I am confident he would have! Here's a curiosity concerning the name WEMBLEY. Using the same alphabet code as we have just used, the sum of the letters in the word WEMBLEY equals 85, which equals (19 plus 66). Wembley was destined for glory in 1966.

O.O'S: That's curious!

Dr. Moogle: Incidentally, the number 11 played a significant role in the 1966 World Cup. The World Cup Finals began on July 11, 1966. That was the 2018th day of the decade. The digits of 2018 sum to 11. The digits of 1966 sum to twice 11. The Finals were played in the 66th year of the century. Note that 66 equals the sum of the numbers from 1 to 11. England's third goal in the World Cup Final was scored by Geoff Hurst in the 11th minute of extra time. The total number of goals scored in the final stages of the Tournament by the eventual champions, England, was 11.

O.O'S: I noticed both England and West Germany had 11 players on their teams.

Dr. Moogle: Your attempt at humor leaves a lot to desire, Mr. O'Shea! I suggest that whatever you may do, do not give up your day job! You will never succeed as a comedian.

O.O'S: I would not like to be a comedian because I am afraid people might laugh at me.

Dr. Moogle: They may laugh at you anyway! As I said earlier, before I was interrupted, the first game of the World Cup Finals in 1966 occurred on July 11. That was the 192nd day of the year. The World Cup Final itself was played on July 30. That was the 291st Saturday of the decade. Note that 291 is the reverse of 192.

O.O'S: Amazing!

Dr. Moogle: The number 192 is related to 1966 as follows: the product of the digits of 192 equals the square root of the product of the digits of 1966.

O.O'S: Astonishing!

Dr. Moogle: The World Cup Final in 1966 was played on the 211th day of the year. Using the earlier alphabet code, the sum of the letters in the words WORLD CUP equals 112, which is the reverse of 211. It is curious that the World Cup Final in 1966 was played on the $(19 + 6 + 6)$ Saturday of the year, exactly $(1 + 9 + 6 + 6)$ weeks before the end of the year and exactly $(1^2 + 9^2 + 6^2 + 6^2)$ days before the end of the year.

O.O'S: Those statistics are amazing!

Dr. Moogle: If you think that that is strange, consider this. Using the earlier alphabet code, A equals 1, B equals 2, C equals 3, and so on, the letters of the three words WORLD CUP FINAL sum to 154. The 1966 World Cup Final was played exactly 154 days before the end of the year.

O.O'S: Holy Moses!

Dr. Moogle: Here is a curiosity that will knock you down. The FIFA World Cup Final Tournament began in 1930. The Final stages of the Tournament were to be played every four years. The Tournaments that would normally have been held in 1942 and 1946 were both canceled due to the Second World War. The Tournament began again in 1950, with the Final stages played every 4 years. Consider this. The World Cup Final in 1966 began at 3:00 p.m. on July 30. That was the 711th hour of the month, on 7/30. It was the 8th World Cup Final in history. The final whistle ending the Final game, which had 30 minutes of extra time, sounded at 5:17 p.m.[1] Now add those four numbers, 711, 730, 8, and 517. The answer is 1966.

O.O'S: Goodness gracious! That is utterly amazing! I note the coincidence that the World Cup Tournament in 1966 began on 7/11 and that the World Cup Final began at the 711th hour of the month.

Dr. Moogle: Yes, indeed. I was wondering if you would spot that.

O.O'S: That's an extraordinary coincidence also!

Dr. Moogle: You obviously did not spot the following curiosity. The final whistle blew at 5:17 p.m. Consider that number. The two outside digits are 5 and 7. Place them together, and one obtains 57, which is the alphabet sum for ENGLAND. The middle digit of 517 is 1. That symbolizes that ENGLAND had just become the number 1 team in the world!

O.O'S: Holy mackerel.

Dr. Moogle: Here is a table that I devised concerning the 1966 World Cup Tournament. Your readers might find it interesting. [The good doctor searched for his notebook. For a few seconds, I did not think he would find it. Eventually, he found it in his inside coat pocket. He took out the tattered old notebook and thumbed through it until he found the page for which he was searching. He tore the relevant page from his notebook, on which the following details were written]:

The 1966 World Cup Tournament

By

Dr. Moogle

First game of Tournament played on 7/11	711
First game of Tournament played on 192nd day of year	192
Final game of Tournament played on 7/30	730
Final game of Tournament played on 211th day of year	211
Duration of soccer game (excluding added time) 90 minutes	90
Total number of games played in 1966 World Cup Tournament	32
Total equals year that the Tournament was played	1966

[I must admit I found the details of this table most interesting.]

O.O'S: What can I say, Dr. Moogle? I must confess to you that I find the contents of that table impressive!

Dr. Moogle: Thank you! Here are five more curiosities that your readers might like concerning the 1966 World Cup. England's captain, the late Bobby Moore, led his victorious team up the famous 39 steps to the Royal Box in Wembley Stadium to collect the World Cup on July 30, 1966. That was a *prime* moment in the life of Bobby Moore. Curiously, the number of *primes* less than 39 is 6 + 6.

O.O'S: The two sixes of the year '66. That's strange. I wonder if any of the England team or members of the media spotted that coincidence.

Dr. Moogle: I don't believe they did. However, I spotted it many years ago. I never published it before, just like the other curiosities I am giving you today. They are all original with me! Here's the second curiosity. As I pointed out earlier, using the usual alphabet code, the sum of the letters in the word ENGLAND is 57. The World Cup Final in 1966 was played on the 211th day of the year, which was 154 days before the end of the year.

The sum of 211 and 154 is obviously 365, which equals the number of days in a year. Curiously, the *difference* between 211 and 154 is 57, which is the alphabet sum for the word ENGLAND.

O.O'S: That's brilliant! What's the third curiosity?

Dr. Moogle: The third curiosity is easily said but not so easy to discover. As I have mentioned, the sum of the letters in the word ENGLAND is 57. England became the number one team in the world when they won the World Cup Final on the 211th day of the year. The sum of the three numbers, 57, 1, and 211, is 269. That happens to be the 57th prime number.

O.O'S: Brilliant! What is the fourth curiosity?

Dr. Moogle: The fourth curiosity is a gem. First, I will point out that the first soccer match ever played at Wembley Stadium was the English Football Association Final on Saturday, April 28, 1923. As mentioned earlier, the World Cup Final in 1966 was held on Saturday, July 30. Every nation at the 1966 World Cup Tournament had named a 22-player squad. From each of these 22-player squads, each nation picked its 11-player team. No substitutes were allowed during any game in the Tournament in England.

O.O'S: I didn't know that!

Dr. Moogle: Well, here's something else I bet you did not know. As I just mentioned, the first soccer match played at Wembley Stadium was on Saturday, April 28, 1923. The World Cup Final on Saturday, July 30, 1966, was played exactly 2,257 weeks after the first-ever match played at Wembley Stadium. Partition the number 2,257 into the two parts 22 and 57. The number 22 symbolizes the 22 players on the English squad in the Tournament in England. The number 57 symbolizes the eventual winner, England, because 57 equals the alphabetical sum of the letters in the word ENGLAND!

O.O'S: That is mind-blowing! Wait until the English media read of these curiosities! Now give me the sixth curiosity.

Dr. Moogle: The sixth curiosity will knock you for six. The World Cup Final on July 30, 1966, was played on the 291st Saturday of the decade. Curiously, if one adds 291 to 1,966, one obtains 2,257.

O.O'S: That is utterly amazing! How do you produce this stuff?

Dr. Moogle: I have an inquiring mind.

O.O'S: Have you any wordplay related to the 1966 World Cup?

Dr. Moogle: Yes, I do. But before I give you wordplay, you should ponder the following curiosity. The alphabet sum for the word ENGLAND is 57. Is it not curious that the only time England won the World Cup Final was in 1966, when the game was played on the 5th Saturday of the 7th month?

O.O'S: That *is* ridiculously crazy!

Dr. Moogle: Glad you like it. You may also find this interesting. The total attendance at all 32 games in the 1966 World Cup Finals in England was 1,614,677.[2] Consider the digits of that number. Note that they sum to 32, the number of games played in the Tournament in England.

O.O'S: Holy smoke! Where do you get these oddities?

Dr. Moogle: I cannot reveal my sources. That is top secret! I am glad you enjoyed that! Now, you asked for some wordplay. Have a look at this card.

[Dr. Moogle took a card from a drawer in the desk in front of him. The following was printed on the card:

A curiosity concerning the 1966 World Cup Final

By

Dr. Moogle

Using the code, A equals 1, B equals 2, C equals 3, and so on, sum the values of all the letters in the paragraph below:

England. Nineteen Sixty-Six: Year when England won the World Cup Final! West Germany equalised in the dying seconds of game. The game went to extra time. England beat West Germany four goals to two in the game.

The sum of all the letters in the above paragraph equals 1966!

I read the card not once, not twice, but three times. I found it a charming curiosity.]

O.O'S: Have you anything else that may interest my readers?

Dr. Moogle: I do! The only man who is credited to date with scoring *three* goals in a World Cup Final is England's Geoff Hurst. The surname HURST is unusual in that it has, in order, *three* consecutive letters of the alphabet. Those *three* letters are the 18th, 19th, and 20th alphabetical letters. Curiously, those numbers sum to 57, which is the alphabet sum for the name ENGLAND.

O.O'S: That's a beautiful curiosity!

Dr. Moogle: I am pleased you like it.

O.O'S: You know, all this talk of the 1966 World Cup Final has reminded me of a little curiosity that has always struck me as odd. There are three main events in English history. The Battle of Hastings in 1066, the Great Fire of London in 1666, and England's World Cup victory in 1966. All three events occurred in years ending in the number 66. What could it mean?

Dr. Moogle: I have often thought about that quirky fact myself. The curiosity may be one of life's great imponderables. Or it may have something to do with the fact that the first letter of the word ENGLAND is the fifth letter of the alphabet. The fifth prime is, of course, 11, and the sum of the numbers from 1 to 11 is 66. If I recall my history lessons correctly, the Battle of Hastings was fought on Saturday, October 14, 1066. Exactly 46,948 weeks later, England won the World Cup at Wembley Stadium, on the $(4 + 6 + 9 + 4 + 8)$st Saturday of the year. The sum of 1,066, 1,666, and 1,966 is 4,698, which is curiously like the number 46,948.

O.O'S: Wonderful!

Dr. Moogle: I am pleased you like the information I am conveying to you.

O.O'S: I most certainly do appreciate it, Dr. Moogle. It is top class!

Dr. Moogle: The difference between 1,966 and 1,066 is 900. When Geoff Hurst scored those three famous goals for England in the World Cup Final in 1966, he was wearing the number 10 shirt. Multiply 900 by 10 to obtain 9,000. Geoff Hurst, who was born on Monday, December 8, 1941, was exactly 9,000 days old on July 30, 1966.

O.O'S: Amazing! That is wonderful!

Dr. Moogle: The day after the World Cup Final was Sunday, July 31. That date may be written as 31/7. It is curious that 317 is the 66th prime number.

O.O'S: That is marvelous! What about the number 1,966? Do you have anything interesting concerning that number?

Dr. Moogle: I do! Let's see. The cube of 1,966 has only digits that are equal to 5 or larger. The number of primes less than 1,966 is 297. That number equals $1 \times 9 - 6 \times 6 + 1 \times 9 \times 6 \times 6$.

O.O'S: Those are the digits of 1,966, twice over!

Dr. Moogle: Precisely!

O.O'S: Have you any curiosity concerning the number 66?

Dr. Moogle: Yes, I certainly do. The number 66 equals $(2 + 5) \times 8 + ((7 - 3)/4) + (1^6 \times 9)$. That expression has all nine digits exactly once. Also, $8 - 7 + 65 = 66 = 4^3 + 2 \times 1$. Note the descending digits on both sides of the equation signs.

O.O'S: I certainly do. Brilliant! I can honestly say that in addition to curiosities and coincidences, you happen to know a thing or two about numbers also.

Dr. Moogle: Thank you! And 66 is the largest number less than 100 that does not have the letter *e*. The next largest number with this property is 2,000.

O.O'S: Wonderful!

With that, my interview with this most strange and curious man ended.

I gathered my documents and notes and bid a long farewell to Dr. Moogle. We shook hands as we parted, and we both promised to stay connected with each other.

I walked out of his office into the next room. His daughter Anna was using this room as a reception area for Dr. Moogle's clients.

I bid farewell to Anna.

I had to walk down a long corridor to reach the elevator. But it was not the long corridor that was on my mind.

No, not at all!

It was the picture of the lovely long, dark hair of Anna that fell gracefully from her beautiful symmetrical face, which was etched on my mind as I left the hotel.

CHAPTER 15

Curiosities Concerning the Gunfight at the O.K. Corral

I awoke at precisely 8:00 a.m. in my bed on a dreary chilly morning in February, wondering what I was going to do for the day. Just then, my cell phone rang. It was Dr. Moogle's secretary, Anna.

"How would you like to come to Tombstone, Arizona," she said, "and experience the feeling of being in a famous place and perhaps meeting a few famous people?"

"That is a very tempting offer," I said. "The mention of the name 'Tombstone' brings back happy memories of my childhood. I remember watching a western series titled *Tombstone Territory* back in the 1960s on Irish television. But who are the famous people that I will be meeting?"

"For starters, me, of course," Anna quipped. "My dad and I decided to come to Tombstone, as we were in this neighborhood. I have squared things with your editor, so your expenses are being looked after. You have no excuses. Get yourself over to Tombstone before the end of the week, and my dad and I will bring you to Boothill Graveyard. There are some famous people buried there. My dad said that if you behave yourself, he may even give you some original material for that column you write in the very forgettable magazine *The Weird and Wonderful*."

"Gosh!" I said. "That will be nice. My editor has asked me a number of times in recent weeks to interview your dad. He told me that I was sure to get new and interesting information."

"Yes, you will, Owen," Anna said. "That is not a problem. My dad will do anything for a quiet life! He said he will give you some info concerning the Gunfight at the O.K. Corral and one or two other things also."

"Great!" I said. "That's wonderful! I am looking forward already to meeting you and your dad in the famous town of Tombstone."

Two days later, I flew into New York and from there to Phoenix, Arizona. Having rested for a day, I traveled by bus from Phoenix to Tombstone. The trip took about three hours.

Before leaving Phoenix, I decided to do a little research on Tombstone. I discovered that the city was founded in 1879. Over a period of three years, the town's mines produced millions of dollars in silver bullion. Tombstone became the county seat of the new Cochise County in 1881. At that time in Tombstone, there was a school, a bowling alley, two banks, three newspapers, and four churches.

But also in Tombstone, there were 14 gambling halls, 110 saloons, many dancing halls, and a large array of brothels. A gang of outlaws known as "The Cowboys" took part in stealing cattle, alcohol, and tobacco from across the Mexican border, which is just 30 miles from Tombstone. Three of these cowboys—Billy Clanton, Tom McLaury, and his brother Frank McLaury—were killed in the shooting incident on Wednesday, October 26, 1881, at 1500 hours, which famously became known as the Gunfight at the O.K. Corral.

The three outlaws were buried in Boothill Graveyard in Tombstone. The men who gunned down the three cowboys were the lawmen Virgil Earp, his brothers Wyatt and Morgan Earp, and Wyatt's friend and associate Doc Holliday. Virgil and Morgan Earp and Doc Holliday were wounded in the gunfight. Two other outlaws, Ike Clanton (Billy Clanton's brother) and Billy Claiborne, who were said to be unarmed at the time, fled the scene of the gunfight and survived. The Gunfight at the O.K. Corral is the most famous gunfight in the Wild West. Incidentally, I also discovered that the letters *O.K.* stand for "Old Kindersley" and refer to the earlier owners of the property.

"Welcome to Tombstone," Dr. Moogle said warmly as I arrived at the Holiday Inn Express, Tombstone. We clasped hands as we greeted each other. Dr. Moogle's daughter, Anna, had told me that she and her dad were staying there, so I had no problem booking in there also.

"This," I said, as I took a pen and notebook from my jacket pocket, "is a famous place in the annals of the American Wild West. Ever since I was a boy back in Cobh, I wanted to visit Tombstone. But I never thought I would get the opportunity. It is a great feeling to be standing here in 'The Town Too Tough to Die.'"

"You obviously did some research on this fine town," Dr. Moogle said. "Tomorrow we will go to Boothill Graveyard. It is in the northwestern part of Tombstone. It gets its name from the fact that most people buried there were outlaws. They mostly died with their boots on. In other words, they died a violent death."

"I would love to visit Boothill. It has been an ambition of mine for a long time to visit this famous cemetery."

"I will show you," Dr. Moogle said, "the graves of Bill Clanton, Frank McLaury, and his brother, Tom McLaury. They are buried side by side in Boothill. There is a grave marker there that states, 'Billy Clanton, Tom McLaury, Frank McLaury. Murdered on the streets of Tombstone, 1881.'"

"I know," I said, "that their killings caused a lot of tension. The Earps and Doc Holliday argued later that they had to kill Billy Clanton and the McLaury brothers, as the outlaws had reached for their guns first. I am also aware that the outlaws had made threats against the Earps and Holliday in the weeks and months leading up to October 26, 1881."

"That's right," Dr. Moogle said. "But in the aftermath of the killings, the townsfolk of Tombstone were divided down the middle on whether the killings were justified. Many believed that the Earps and Doc Holliday had no choice but to kill the outlaws. But just as many townsfolk in this old town believed that the killings of Billy Clanton and the two McLaurys was cold-blooded murder perpetrated in broad daylight on the streets of Tombstone."

"I see," I said as I took note of what Dr. Moogle was saying.

"In any event," Dr. Moogle said, "the gunfight in Tombstone is now probably the most famous gunfight of the Old West. Before I came here, I decided to do a little research on the famous Gunfight at the O.K. Corral. I discovered a few curiosities. The readers of your column in *The Weird and Wonderful* might like to read of them. But I will give you them tomorrow. First, you need a good night's rest. You must be tired after all your traveling."

The next day, Dr. Moogle and I saw a daily reenactment of the famous and historic gunfight close to the O.K. Corral. We also visited Boothill Graveyard. We went to the graves of Billy Clanton and Frank and Tom McLaury.

A few minutes later, Dr. Moogle showed me the grave of one Lester Moore. He had been killed in or around 1880. Moore had had a quarrel with one Hank Dunstan. The quarrel escalated, and both men drew their guns. Dunstan fired four shots from his .44-caliber revolver, hitting Moore in the chest with each of the four bullets. Moore fired just one shot at Dunstan, fatally wounding him. Les Moore died from his wounds. His tombstone epitaph reads, "Here lies Lester Moore. Four slugs from a 44. No Les, No More."

"That's a fitting epitaph," I said, pointing to the inscription over Moore's grave, "if ever I saw one."

We then took the seven-minute walk back to the Holiday Inn Express in Tombstone. We went to one of the nonsmoking rooms, where we made ourselves comfortable and ordered some cool beers.

"Here is some trivia that your readers might like," Dr. Moogle said as he rubbed his well-trimmed beard between his forefinger and thumb. "You are aware that a blockbuster Hollywood movie was made of the famous gunfight in Tombstone? The movie was called *Gunfight at the O.K. Corral*. It was distributed by Paramount Pictures and was released in May 1957. The well-known actor DeForest Kelley played the part of Morgan Earp in the movie. Two years earlier, in 1955, DeForest Kelley played the part of Ike Clanton in the *You Are There* episode titled 'Gunfight at the O.K. Corral.' Curiously, in 1968, DeForest Kelley played the part of Chief Medical Doctor Leonard McCoy in the *Star Trek* episode titled 'Spectre of the Gun.' In this episode, Captain Kirk, Mr. Spock, Doctor McCoy, Scotty, and Chekov are forced to reenact the famous gunfight, playing the part of the outlaws in the famous showdown. Doctor Leonard McCoy is forced to play the part of Tom McLaury, one of the three men killed in the gunfight. Thus, at one time or another, DeForest Kelley played the role of three of the characters in the famous gun battle."

I took notes to help recall what the good doctor was saying.

Dr. Moogle reached for his wallet and took a beautiful laminated card from it. The card had a colored picture of a cowboy wearing guns and holsters up in the left-hand corner. The following was written on the card:

A Curiosity Concerning the Gunfight at the O.K. Corral

By
Dr. Moogle

The grisly gunfight at O.K. Corral occurred at fifteen hundred hours on Wednesday, October Twenty-Sixth, Eighteen Eighty-One. The gunfight lasted thirty seconds. Thirty rounds were fired.

Using the alphabet code where A equals 1, B equals 2, C equals 3, and so on, the sum of the letters in the three sentences above equals 1881.

The Gunfight at the O.K. Corral occurred in 1881.

I read the contents of the card. "That *is* extraordinary," I said. "You never lost your touch for finding curiosities, Dr. Moogle. May I begin the interview with you now?"

"You certainly may." I took out my pencil and notepad and began taking notes. The following interview then took shape.

O.O'S: How did you first get interested in the Gunfight at the O.K. Corral?

Dr. Moogle: I have been interested in the stories and heroes of the Wild West ever since I was a little boy. I recall reading when I was about 10 years old that the Gunfight at the O.K. Corral occurred at about 1500 hours, extremely near 312 Fremont Street, in Tombstone, in a vacant lot west of Fly's Photography Store. I never forgot the details. Here is a card that I prepared the other day. Your readers may find the contents of it interesting:

A Curiosity Concerning the Gunfight at the OK Corral

By

Dr. Moogle

The Gunfight at the O.K. Corral began at 1500 hours	1500
The Gunfight occurred near 312 Fremont Street	312
The Gunfight lasted 30 seconds	30
It is believed 30 rounds were fired in the gunfight	30
The Earps' party consisted of 4 men	4
The McLaury - Clanton party suffered 3 fatalities	3
Two men, Ike Clanton, and Billy Claiborne, fled the scene	2
Total equals the year in which the gunfight was fought	1881

O.O'S: Wonders will never cease! I like it! I think my readers will like these details.

Dr. Moogle: Here is something else that your readers might like. Consider the year of the gunfight. It was 1881. Partition that number into the two parts 18 and 81. The product of those two numbers is 1,458. It was close to 14:58 p.m. on the fateful day when the men walked to the O.K. Corral to face each other. Now 1,881 minus 1,458 equals 423. Remember that number! Consider the names of the three outlaws who were killed that day in Tombstone. They were Billy Clanton, Tom McLaury, and Frank

McLaury. Using the usual alphabet code where A equals 1, B equals 2, and so on, the sum of the letters in the three men's names is 423!

O.O'S: Holy cow! That *is* curious.

Dr. Moogle: Glad you like it. The alphabet sum of the two outlaws who fled the scene, Ike Clanton and Billy Claiborne, is 243. That number has the same digits of the number 423.

O.O'S: Suffering catfish! That's most interesting.

Dr. Moogle: I am pleased you feel that way! The number 1,881 is an interesting number. It equals $2^7 + 3^6 + 4^5$. Note the order of the digits! The number of primes less than 1,881 is $(1 \times 8 + 8 + 1)^2$. And the number 423, which is connected to the names of the three outlaws, is interesting also. The number of primes less than 423 is $1^8 + 81$. Of course, if you square the sum of the digits in 1,881, you obtain 324. Reverse that number, and you get 423.

O.O'S: I'll be darned. I think the folk living here in Tombstone would love to read of these curiosities. Have you anything else?

Dr. Moogle: Well, it's funny you should say that! There are many people in Tombstone who will tell you that there was an evil act conducted here on that fateful day in October 1881. Of course, it should come as no surprise to those people that 423 plus 243 equals the *number of the beast*, 666.

O.O'S: That is utterly amazing!

Dr. Moogle: You may not know that Doc Holliday was a dental surgeon. However, the Doc gave up the practice after a brief time, as he was able to make more money by gambling. He was born John Henry Holliday on August 14, 1851. He died of consumption at the age of 36 on November 8, 1887. Holliday was a great friend of Wyatt Earp. Wyatt argued that Doc Holliday saved his life one day in a saloon when desperadoes surrounded Wyatt. Here is a little curiosity concerning Doc Holliday. The Doc was born on the $1 + 3 + 2 + 3 + 5$th day of the month and lived for 13,235 days!

O.O'S: That *is* amazing! I have often read about the Doc, but I never read that curiosity before.

Dr. Moogle: There is an explanation for that. I never published these curiosities before. They are all original with me.

With that, the interview ended, at least for a while. Dr. Moogle said that he was feeling tired and that he would like to take a nap for an hour or so.

Anna and I went for a walk around Tombstone. It was a chance for me to get to know her a little better.

"Are you also into the events of the Wild West?" I asked.

"Well, I suppose I am. My dad has always told me interesting stories about the Wild West when I was growing up here in the United States. I grew up in New York."

"The city that never sleeps!" I said, proud that I could remember that saying about the Big Apple.

"Yes," Anna said, "it is a fast town. In my opinion, New York is not a city. It's a world! That is a philosophical way of expressing one's view about New York."

"I like it," I said. We walked a little further without saying a word. Then I broke the silence. "Were you always interested in numbers, Anna?"

"I was. Gosh, I can remember when I was nine years old, I pointed out to a friend that I was named Anna because the alphabet sum of my name equals 30. I was born on the 30th day of the year. I also pointed out that 30 is the sum of the first four squares. That accounted for the first four integers in order! When I gave her that information, she thought I was some kind of freak! Then I told her that my birthday in a non–leap year arrived just 5 times 67 days before the end of the year. That accounted for the next three digits in order! My friend was stunned. She went home and told her mom, who then thought I was some kind of freak but in a wonderful way, I think! At least I hope so! I think you may have noticed that I happen to love numbers, Owen."

"That's great! I love numbers too, Anna."

That night, the three of us went to Big Nose Kate's Saloon in Tombstone for food and drinks. The bar is named after Doc Holliday's girlfriend. Dr. Moogle told me that Big Nose Kate was apparently "the broad who loved Doc Holliday but who loved every other cutthroat, murderer, and desperado too." Big Nose Kate's Saloon was originally known as the Grand Hotel. It was built in 1880. Ike Clanton and the two McLaury boys stayed there on October 25, 1881, the night before the world-famous Gunfight at the O.K. Corral.

It was not long before the subject turned to other famous outlaws of the Wild West. Thus, in Big Nose Kate's Saloon, the second part of the interview took place.

O.O'S: Have you anything further to say about the Wild West, Dr. Moogle?

Dr. Moogle: Plenty! Of course, any discussion of the Wild West must include the life of one of the most famous outlaws of that period, namely, Jesse James. He was born, you know, on Sunday, September 5, 1847. That date is usually written in the United States as 09/05/1847. Jesse was murdered by Robert Ford, one of his own men, on Monday, April 3, 1882. Here is a little curiosity about Jesse that I discovered and never published before. Jesse was $0 + 9 + 0 + 5 + 1 + 8 + 4 + 7$ years old when he was killed.

O.O'S: Holy dynamite! Those are the digits of his date of birth! That is almost frightening!

Dr. Moogle: Don't get too scared! It ain't good for your health. Here is something else that I hope will not frighten you or your readers. Hopefully, your readers might like what I am about to say. Jesse James, whose father was Irish, was born on the 1st day of the week, and he was killed on the 2nd day of the week, when he was 34 years old; Jesse was killed on the $5 + 6 - 7 + 89$th day of the year.

O.O'S: Those are the nine digits in order. Amazing!

Dr. Moogle: I had a feeling you would like that! Here is a curiosity my daughter, Anna, discovered concerning Jesse Woodson James, to give the famous outlaw his full name. The initials of his name are the 10th, 23rd, and 10th letters of the alphabet. Those three integers sum to 43. Jesse Woodson

James was killed on 4/3 in 1882. My daughter is very clever to have spotted that curiosity.

O.O'S: I agree entirely! That is an astonishing curiosity! I would never have spotted that coincidence! Your daughter is extremely good at spotting these oddities. I did, however, spot the following little coincidence only very recently. One of Jesse's men, Robert Ford, shot Jesse dead. Using the earlier alphabet code, the sum of the letters in the name FORD equals 43. Jesse was killed on 4/3.

Dr. Moogle: Very neat, Mr. O'Shea. You are getting good at this game! There is hope for you yet! All this talk of the Wild West brings to mind one of the youngest and most famous outlaws of those far-off days. I speak of a young man who is still—after all this time—a legend of the Wild West. The man I speak of is none other than the famous Billy the Kid. His parents were Irish, you know. Historians of the Old West believe that this famous shooter was born William Henry McCarty Jr. in New York on November 23, 1859. The number 7, or multiples of 7, appeared at significant times in the life of Billy the Kid.

O.O'S: You're kidding. Oops, I did not intend the pun!

Dr. Moogle: I ain't kidding mister! There are 7 letters in the words NEW YORK. There are 21 letters in the name WILLIAM HENRY MCCARTY JR. Of course, 21 is three times 7. William was aged 14 when his mother died; 14 is twice 7. William changed his name to William H. Bonney. This name has 14 letters. According to legend, Billy the Kid killed 21 men, though some historians think that the more likely figure is 7. Billy the Kid was shot dead by Sheriff Pat Garrett on July 14, 1881. July is the 7th month. The surname GARRETT has 7 letters, and the surname begins with the 7th letter. Billy the Kid is laid to rest in Old Fort Sumner Cemetery in New Mexico. The phrase OLD FORT SUMNER CEM-ETERY has 21 letters. The first letter of New Mexico is the 14th letter of the alphabet. Curiously, Billy the Kid was 21 years, 7 months, and 21 days old when he was killed. These are just some of the places where 7, or multiples of 7, appear in the life of Billy the Kid.

O.O'S: Astonishing! That is wonderful! It beats me how you produce this mate-rial. I am greatly confident my readers will enjoy reading these curious facts. Moving the subject on, I was wondering if you have any curi-ous information concerning movie stars my readers might enjoy. For example, I remember in 2012 that I was saddened to read of the death of television actor Larry Hagman. He was well known for the part of J.R. that he played in the TV hit *Dallas*. That show was very popular in Ireland.

Dr. Moogle: Yes, indeed! "Who shot J.R.?" was a big question here in the United States in 1980. That was surprising given the fact that in the real world, there was a lot happening. The American presidential race was on. At the time, American hostages were being held in captivity in Iran. The Iraq-Iran

War was raging. Then, of course, there was the Polish Revolution. But most people around the world had only one thing on their mind: Who shot J.R. Ewing on March 21, 1980? The episode that revealed who the shooter was aired on November 21, 1980.

O.O'S: Yes, I agree. It is amazing that the world was more concerned with a fictional story than what was happening in real life. We humans are a strange species!

Dr. Moogle: Yes, we are, and some members of that species are very strange indeed! Now here's the thing. The *Dallas* episode that revealed the shooter of J.R. first aired in the United States on November 21, 1980. President John F. Kennedy was assassinated in Dallas on November 22, 1963. And Larry Hagman died in Dallas on November 23, 2012. One has three dates here: November 21, 1980; November 22, 1963; and November 23, 2012. Curiously, all three of those dates fell on a Friday.

O.O'S: Amazing! Thank you for all this curious information, Dr. Moogle. My readers will like it, I think. But I know my readers also like a little bit of mathematics. Have you anything interesting that I can give them?

"I can give them some interesting information," Anna said as she joined the conversation. "It concerns the sums of infinite series. Suppose, for example, one is asked to sum the following series: $1/2 + 1/4 + 1/8 + \ldots$ and so on. Here is an effortless way to go about it. Let x equal the sum of the series. Then $2x$ equals $1 + 1/2 + 1/4 + 1/8 + 1/16 + 1/32 + \ldots$. From the second term on, the second series is the same as the first series. Thus, we can write this." Anna took a page from *my* notepad and wrote down the following:

$$x = \qquad 1/2 + 1/4 + 1/8 + 1/16 + 1/32 + \ldots$$

$$2x = 1/1 + 1/2 + 1/4 + 1/8 + 1/16 + 1/32 \ldots$$

Subtract x from $2x$.

Thus $x = 1$.

"That is a very clever method," I said. "Does it always work?"

"It works for all geometrical series," Anna said. "Geometrical series are those series where the ratio between successive terms is equal. There is another effortless way to sum the series that I have just mentioned. Consider again the series $1/2 + 1/4 + 1/8 + 1/16 \ldots$ and so on. To see what the sum of this series equals, consider the following. Suppose you have one bar of chocolate. That bar of chocolate consists of 1/2 of a bar of chocolate, 1/4 of a bar of chocolate, 1/8 of a bar of chocolate, and so on, each segment being just half the size of the earlier segment. When one adds *all* the segments of the bar of chocolate, the result has to equal one bar of chocolate. The key thing here is that we add *all* the members of the series. In other words, the sum of *all* the members of the series $1/2 + 1/4 + 1/8 + 1/16 + \ldots$ and so on equals 1."

"That's wonderful, Anna," I said. "That analogy of the bar of chocolate makes it clear how the sum of *all* the terms of that particular series sums to 1."

"Glad you like it," Anna said.

"Consider the following infinite series," Dr. Moogle said. The good doctor took out his notepad and wrote the following down:

$$(1/1 - 1/2) + (1/2 - 1/3) + (1/3 - 1/4) + (1/4 - 1/5) + \ldots + (1/(n) - 1/(n + 1))$$

"The trick here," Dr. Moogle said, "is to note that from the third term on, each term cancels with the earlier term. Thus, negative 1/2 and plus 1/2 cancels each other; negative 1/3 and plus 1/3 cancels each other, and so on. As we sum an increasing number of terms, all of the terms cancel each other, except the first term and the last term. As the number of terms in the series increases toward infinity, the value of the "last" fraction approaches zero. Thus, the *limiting sum* of the series equals 1 + 0, or 1."

I was scribbling all this down at a frantic rate. "That is truly wonderful," I said.

"I am pleased you like it," Dr. Moogle said. "Here is an interesting fact. Consider the following series: 1/1 + 1/3 + 1/6 + 1/10 + 1/15 + . . . and so on. That series is the sum of the reciprocals of the triangular numbers. Obviously, the first term equals 1; the sum of the next two terms equal 1/2; the sum of the next four terms equal 1/4, and so on. We know from what Anna has said earlier that the limiting sum of the series 1/2 + 1/4 + 1/8 + . . . is 1. Thus, the limiting sum of 1 + 1/2 + 1/4 + 1/8 + . . . is exactly 2. Since this latter series, 1 + 1/2 + 1/4 + 1/8 + . . . , equals 1/1 + 1/3 + 1/6 + 1/10 + 1/15 . . . , we know that the limiting sum of the series of the reciprocals of the triangular numbers is exactly 2."

"Gosh!" I said. "That is a beautiful result! It is so elegant."

"When one thinks about it," Dr. Moogle said, "it is amazing that one can add an infinite number of fractions and find that the result is an exact integer!"

"Yes, indeed it is!" I replied.

"As you may know," Dr. Moogle said, "geometric series are used throughout mathematics. They are also widely used in physics and biology. Here is a handy trick that your readers might like. Consider a famous date in history, say, 9/11. If one divides 911 by 999, one gets 0.911911911911. . . , where the three-digit number 911 infinitely repeats. One may wonder how to obtain such a fraction that gives an infinite decimal repeat of the desired integer. The trick here is to let x equal 0.911911911. . . . Then 1,000x equals 911.911911911. . . . Subtract x from 1,000x. The result tells us that 999x equals 911. Therefore, x equals 911/999, which gives the desired repeating decimal."

"That's brilliant," I said. "But I suppose it is fair to say that the average schoolchild does not come across infinite series in his or her mathematics studies."

"Yes, they do," Dr. Moogle said. "Consider the recurring decimal 0.3333333. . . . It is equal to 1 divided by 3. The recurring decimal tells us that 1/3 equals 3/10 + 3/100 + 3/1,000 + 3/10,000 + . . . and so on. In other words, 1/3 is equal to the sum of an infinite number of fractions. Similarly, 1/9 equals 0.111111. . . . Another way of expressing this is to write 1/9 and point out that it equals 1/10 + 1/100 + 1/1,000 + 1/10,000 + . . . and so on. This leads to the conclusion that

9/9, or 1, equals 9/10 + 9/100 + 9/1,000 + 9/10,000 + . . . and so on. Thus, 1 equals 0.999999. . . . This is a solid result, obtained with ironclad logic, which many non-mathematicians find hard to accept. But the result is true."

"Amazing!"

"Steven Strogatz," Dr. Moogle said, "the American mathematician born in 1959 and thankfully still with us, is on record for saying that 1 equals 0.999999. . . is his very favorite equation.[1] Strogatz says that the left-hand side of the equation, 1, is the smallest integer in math and is the beginning of math, while the right-hand side of the equation, 0.999999. . . , has all the mysteries of infinite series."

"Of course," Dr. Moogle said, "both e and pi are transcendental numbers, which are each the limit of an infinite series. Here is an infinite series that your readers might like to sum. With a little insight, it is ridiculously easy to solve. Find the sum of 3 + 5/2 + 7/4 + 9/8 + 11/16 + 13/32 + 15/64 + ..."

"I will give it to my readers," I said. The solution is given at the end of this chapter.

"Try your readers with this one," Anna said. "Find the sum of the following infinite series: (1/2 − 1/4) + (1/3 − 1/5) + (1/4 − 1/6) + (1/5 − 1/7) + (1/6 − 1 / 8) + (1/7 − 1/9) + . . . and so on."

"I'll give that to my readers also," I said.

"Here is a famous problem," Dr. Moogle said, "that can be solved using infinite series. However, there is a far easier approach in solving the problem. Let us assume that there is a country in which all couples continue to have children until they have a boy. If the firstborn child is a boy, then that couple will not have any more children. If the firstborn child is a girl, then the couple will have a second child, a third child, or whatever until they have a boy. Once a couple have a boy, they do not produce any more children. What is the proportion of boys to girls in that country? Let us assume that in this country no one dies and that the probability of having a boy is 50 percent."

"I do not have a clue," I said. "I think I might be biting off more than I can chew with that puzzle."

"Don't worry," the professor said, "about biting off more than you can chew; your mouth is probably a whole lot bigger than you think."

"Wait a minute!" I said. "I think I have just had a brainstorm. Surely, there will be a lot more girls than boys in the population in that country."

"Don't rush to solutions, Owen," Anna said. "Give the problem some time. Look at it this way. The probability that the first child, let us call it B, born to a couple will be a boy is 50 percent, or 1/2. Of two children born to a couple, the probability that the second will be a boy, GB, is 25 percent, or 1/4. The probability that of three children born to a couple, GGB, the third child will be a boy is 12½ percent, or 1/8. Therefore, the probability that the total number of boys born is 1/2 + 1/4 + 1/8 + 1/16 + ... This series, as we have seen, has an infinite sum equal to 1. Of course, the same principle applies to girls. The probability that the total number of girls born is 1/2 + 1/4 + 1/8 + 1/16 + . . . , which also sums to 1. Therefore, the ratio of boys to girls is 1 to 1. The proportion is therefore equal."

"I think I'm with you," I said.

"Congratulations!" Dr. Moogle said. "At long last, you are using some of the gray matter between your ears! We must be thankful for that! However, there is a simpler way of approaching the children problem. One does not need to sum an infinite series to solve this little teaser. Here's the easy solution. Imagine you are the census taker in the relevant country. Let us assume that every birth is recorded and that the details are sent to your office. For each birth, there is a 50 percent chance that the child will be a boy and a 50 percent chance that the child will be a girl. No matter what social convention exists in the country, the ratio for boys to girls must be 50 to 50, or 1 to 1."

"I'm with you," I said. "At least I think I'm with you. Ah, I have it now! Thank you both for the interesting info on infinite series. Thank you both also for the interesting info on the cowboys of the Wild West. Now I think I will spend some good old-fashioned dollars in this famous old town. I earned a lot of dollars recently, and I think I should dispense some of them and have a good old time in the company of two particularly good friends."

"That's not a bad idea," Dr. Moogle said. "But just remember what a wise old cowboy once said: 'Never let your yearnings get ahead of your earnings.'"

Solutions

1. Dr. Moogle asked the reader to find the sum of $3 + 5/2 + 7/4 + 9/8 + 11/16 + 13/32 + 15/64 + \dots$.

    Let $x = 3 + 5/2 + 7/4 + 9/8 + 11/16 + 13/32 + 15/64 + \dots$.

    Then $1/2x = 3/2 + 5/4 + 7/8 + 9/16 + 11/32 + 13/64 + \dots$.

    Subtract $1/2\,x$ from x. This gives

    $1/2x = 3 + 2/2 + 2/4 + 2/8 + 2/16 + 2/32 + 2/64 + \dots$

    $1/2x = 3 + 1 + (1/2 + 1/4 + 1/8 + 1/16 + 1/64 + \dots)$

    You will note that the series in parentheses is our old friend $1/2 + 1/4 + 1/8 + 1/16 + 1/64 + \dots$. We have already seen that this series sums to 1. Therefore, we can write

    $$1/2x = 3 + 1 + (1/2 + 1/4 + 1/8 + 1/16 + 1/64 + \dots)$$
    $$= 1/2x = 3 + 1 + 1$$
    $$= 1/2x = 5$$

    Thus, $x = 10$.

    Thus, the limiting sum of the series $3 + 5/2 + 7/4 + 9/8 + 11/16 + 13/32 + 15/64 + \dots$ is 10.

2. Anna asked the reader to calculate the sum of $(1/2 - 1/4) + (1/3 - 1/5) + (1/4 - 1/6) + (1/5 - 1/7) + (1/6 - 1/8) + (1/7 - 1/9) + \dots$.

    The reader will note that as the number of terms approaches infinity, all the terms of the series, except 1/2 and 1/3, cancel each other. Thus, the limiting sum of the series is simply the sum of the first two terms: $1/2 + 1/3$, or $5/6$.

CHAPTER 16

A Curious E-Mail from Dr. Moogle

I received a curious e-mail from Dr. Moogle about one month after I had said good-bye to him and his lovely daughter, Anna, in Tombstone.

The following is the e-mail:

To Mr. Owen O'Shea.

Did you know that Salt Lake City is the home of one of the most famous people that lived on this planet?

I write of the mega-savant Kim Peek. He was born in Salt Lake on November 11, in 1951.

Kim Peek was a remarkable man. He had an incredible photographic memory. Kim read about 13,000 books in his lifetime. He was able to remember about 98 percent of everything he had read. He particularly liked doing the calendar trick, that is, naming the day of the week for any date called out by a spectator.

As readers may know, Dustin Hoffman played the part of the central character, Raymond Babbitt, in the movie *Rain Man*. The character Raymond Babbitt was based on Kim Peek.

Kim loved numbers. This fact is reflected in his name and in his date of birth. The initial letter of his first name and the last letter of his surname is the 11th letter of the alphabet. Kim was born on 11/11 in 1951. Also, the initial letter of his surname is the 11th letter from the *end* of the alphabet. The number 1,951 is not evenly divisible by 11. However, 1,951 is the 297th prime, and 297 *is* evenly divisible by 11. When 1,951 is divided by the sum of its digits, the integer part of the answer is equal to 11^2.

Using the alphabet code where A equals 1, B equals 2, C equals 3, and so on, the sum of the letters in the seven-letter name KIM PEEK is 70. Using the same code, the sum of the letters in the words RAIN MAN is also 70. This apparently illustrates to all and sundry that Kim was the genuine RAIN MAN. Kim's father, Fran Peek, wrote a book titled *Kim Peek: The Real Rain Man*. As I said earlier, Kim was born in 1951. Of course, 70 equals 19 plus 51. By the way, the seven letters of the name KIM PEEK, plus 70, gives 77. Using the same code, the sum of the letters in the word SAVANT is 77.

There were two years in Kim Peek's life that proved highly significant to Kim. Those years were 1984 and 1988. In 1984, Kim Peek met movie screenwriter Barry Morrow by chance in Arlington, Texas. Morrow was so taken by Peek's memory and abilities that he wrote the script for *Rain Man* and later publicly acknowledged that that famous movie was inspired by Kim. The movie changed Kim's life forever. Kim reached the age of 33 in 1984. Using the earlier alphabet code, the sum of the letters in the word KIM is 33. In 1988, the movie *Rain Man* was made. In that year, Peek reached the age of 37. Using the earlier alphabet code, the sum of the letters in the word PEEK is 37.

In 1962, Kim Peek reached his 11th birthday on 11/11. Kim was then exactly 4,018 days old. Note that $4 + 0 - 1 + 8$ equals 11. That time period is exactly 574 weeks. Curiously, 574 equals $123 - 4 - 567 - 89 + 1,111$.

Kim Peek was born in 1951 and died in 2009. There is a pattern to his year of birth and his year of death. Add the sum of the digits of 1,951 to 1,951 to obtain 1,967. Add the sum of the digits of 1,967 to 1,967 to obtain 1,990. Finally, add the sum of the digits of 1,990 to 1,990 to obtain 2,009.

The numbers 1,951 and 2,009 can be expressed in a somewhat similar manner:

$$1,951 \text{ equals } 1 \times 2 + 34 \times 56 + 7 + 8 + 9 + 10 + 11$$
$$\text{and}$$
$$2,009 \text{ equals } 1 \times 2 + 34 \times 56 - 7 + 89 + 10 + 11$$

Kim Peek was a walking encyclopedia. His mind could accurately be described as being a great *store* of knowledge. Of course, when the word STORE is used as a verb, it means "to keep." The word KEEP is Kim's surname in reverse.

Using the alphabet code I gave earlier, the sum of the letters in the phrase I: KIM PEEK: MALE GENIUS is 185, and the sum of the letters in the word SAVANT is 77. Concatenate those two numbers to form the number 18,577. Kim Peek was born on the 18,577th day of the century.

The sum of the letters in the word SAVANT is 77. Kim's surname consists of four letters. If one multiplies 77 by 4, one obtains 308. Reverse the digits of that number and add it to 308. The answer is 1,111. Kim was born on 11/11.

Kind regards,
Dr. Moogle

CHAPTER 17

O'Mara, the Mathematical Seafarer

One morning in early March 2021, the editor of the magazine *The Weird and Wonderful*, in which I write a monthly column, phoned me and told me that he wanted me to interview a certain individual named Jack O'Mara, who had moved recently to Dublin, Ireland.

"O'Mara is a rather interesting character," my editor said. "A good friend in the newspaper business recommended him. He works four months at a time at sea, then he is off for two months. O'Mara likes numbers and curiosities of all sorts. I think he may have information that you can use in your monthly column. I expect you to interview O'Mara within the next two weeks. I will send you an e-mail with his photograph and contact number for you to set up a meeting with him. Incidentally, he is used to people addressing him as simply 'Mara.'"

Two weeks later, I went to Dublin to meet Mr. O'Mara. I found him in a lounge bar, in a well-known city-center hotel, drinking from an ice-cool beer. Mara was aged about 30. He was wearing blue jeans and a leather bomber jacket, and white sneakers. Mara was tall and thin, had piercing blue eyes, was clean shaven, and had a fine crop of fair hair on his head.

I ordered a nonalcoholic drink and joined this friendly-looking stranger.

"Hi there," I said. "I'm Owen."

"I'm Mara," he said. "It's nice to meet you."

"It's nice to meet you too!"

We made small talk for 5 or 10 minutes before I finally got around to the reason for my meeting with him.

"I hear you are interested in all sorts of curiosities," I said matter-of-factly as I took my pen and notepad from my pocket.

"That's correct," Mara said. "I like the strange and the unusual. That is why I subscribe to *The Weird and Wonderful*. I usually enjoy your monthly articles. I often bring them to sea with me to read them when I am off duty. I recall reading last month's article. From the moment I put it down, I could not pick it up again."

"It is very nice of you to say so," I said. "How did you get interested in the subject of the strange and the wonderful?"

"Well," Mara said. "I have met many strange and weird characters over the last 15 years or so. They seem to come out of the woodwork when one is not looking! They usually look normal. But one should not be fooled by that! But that aside, I have been interested in the strange and wonderful since I was a young boy. For instance, I recall when I was a boy of about eight, I was always looking out for interesting postage stamps. I discovered that Monaco issued a 50-cent stamp in 1947, depicting President Franklin D. Roosevelt working on his stamp collection. In the depiction, Roosevelt has six fingers on his left hand! I also discovered that a U.S. 3-cent stamp issued in 1946 to commemorate the 100th Anniversary of the Entry into Santa Fe shows two horses side by side, but the two horses have only seven legs between them!"

"Goodness gracious!"

"Or consider the fact," Mara said, "that the U.S. Postal Service issued a 24-cent stamp in 1918 that is called an "Inverted Jenny." It got its name because it depicts an inverted Curtiss JN-4 airplane in the center of the design. There were many correct Jenny stamps issued that year, but one sheet of 100 of these stamps showing the airplane upside down were printed. It is believed that only 94 of those Inverted Jenny stamps exist today. They are extremely valuable. One such stamp sold in 2018 for over $1.5 million."

"Very nice indeed," I said. "Curiously, in 2019, the Irish postal service issued a £1 stamp to commemorate the first moon landing in July 1969. The Irish word *Gaelach*, meaning 'Irish,' appeared on the stamp, instead of the Irish word *Gealach*, meaning 'moon.'"

"Yes, I recall that error," Mara said. "I can see you like the strange and the curious too."

"I have to admit that I do," I said. "I suppose that being a seafarer, you hear lots of stories about strange events."

"I do," Mara said, "but not all of them are true! I will give you some stories today, but all of them are authentic. Being a seafarer has taught me many things that land lovers may have missed. For instance, when I was young, I believed money was the most important thing in life; now that I'm older, I know it is!"

"I take your word on it! Do you mind me asking if you have a wife and family?"

"No, I don't," Mara answered honestly. "Only last month, one of my shipmates, at the age of 21, got married. Believe it or not, I made a pledge to myself when I was 21 to never get married young, even if I live to be as old as Methuselah!"

"That was very sensible of you," I said, reassuringly.

Mara took another sip at his cool beer. As he raised his glass, he quipped, "As W. C. Fields (1880–1946) once said, 'I never drink water; that's the stuff that rusts pipes.' I could not agree more with Fields."

"I am not sure I can agree with you on Fields's point of view," I said.

"Not to worry," Mara said. "Let's agree to disagree."

"Good idea!" I said.

"Where do I begin!" Mara said. "Okay, let's see. I have always tried to follow events in the Emerald Isle closely, even when I was far from its shores. Did you know that the 2011 election for the president of Ireland was unusual? There were 7 candidates running in the election, all aiming to become Ireland's 9th president in the

11th year of the century. The election was held on the 27th of October. Is it not curious that $7 + 9 + 11$ equals 27?"

"I suppose it is!" I said.

"I also find it strange," Mara said, "that the Irish president was sworn in on 11/11/11. Here's why. Using the usual alphabet code where A equals 1, B equals 2, C equals 3, and so on, the sum of the letters in the word PRESIDENT equals 110, which equals 11 times 11 minus 11."

"Nice one," I said. "I would never have spotted that."

"I didn't think you would," Mara said. "There's more, however. Using the same alphabet code, the sum of the letters in the word IRELAND and the sum of the letters in the word IRISH equal 63. In binary code, the number 63 equals 111111."

"Amazing!" I said.

"President Higgins," Mara said, "was sworn in on 11/11/2011. Consider the following multiplication and addition made with the digits of that date: 11 times 11 plus 20 plus 1 plus 1. The result is 143, which equals $2^2 + 5^2 + 7^2 + 7^2 + 4^2$. Michael D. Higgins, who was born on April 18, 1941, was 25,774 days old on the day he was sworn in as president of Ireland."

"Amazing!" I said. "I never read that before!"

"That is because it is original with me, and I never published it before!"

I jotted down the details. "There are many curious coincidences," I said, "involving numbers and words. I will give you a few of my own little discoveries. Using the usual code, A equals 1, B equals 2, C equals 3, and so on, the sum of the letters in the word SQUARE equals 81, which is, of course, a square number. The sum of the letters in the two words SQUARE DIGITS equals 149, and each of those digits, 1, 4, and 9, is square."

"I didn't know those curios," Mara confessed. "Did you know that the five vowels, *a, e, i, o,* and *u,* all fall at *odd* positions within the alphabet? Also, every odd number contains the letter *e.* Using the earlier alphabet code, the letters of the alphabet sum to 351, a number having the first three odd digits. Since there are 26 letters in the alphabet, each letter has an average value of 13.5, a number also having the first three odd digits. Using the earlier code, the sum of the letters in the phrase CALL TWENTY SIX LETTERS 'ALPHABET' equals 351."

"By heck," I said, "that's odd."

"Consider the fact," Mara said, "that the alphabet sum of the four directions, north, south, east, and west, is 270. Of course, those are the four cardinal points on a compass. It is curious that the alphabet sum of the words in the phrase CARDINAL POINTS ON COMPASS is also 270."

"Gosh! That is amazing!

"I am pleased you like it," Mara said. "I presume you know of William Rowan Hamilton."

"I certainly do. I know he was born in Dublin in August 1805 and died in September 1865."

"That's correct," Mara said. "Hamilton was one of Ireland's greatest scientists. In addition, he was also an excellent mental calculator."

"I didn't know that," I said surprisingly.

"Here's something else you probably don't know," Mara said. "It's a curiosity concerning Hamilton that I never before published. It involves the number 4, or multiples of 4. Hamilton was born on the 4th day of the month. He was born in Dublin. The first letter of DUBLIN is the 4th letter of the alphabet. The first letter of HAMILTON is the 4th plus 4th letter of the alphabet. He was the 4th eldest child in his family. The number of letters in the name WILLIAM ROWAN HAMILTON is 4 squared plus 4. It was on the 4 squared day of October in 1843 that Hamilton discovered his famous formula involving quaternions. The sum of the digits in 1,843 equals 4 squared. October is the 10th month. Of course, 10 is the sum of the integers from 1 to 4. Hamilton died on September 2, 1865. The number 2 is the square root of 4. The digits of his year of death sum to 4 squared plus 4. Hamilton was aged 60 when he died. Of course, 60 equals 44 plus the square root of 4 raised to the power of 4."

"Holy cow!" I exclaimed. "That is extraordinary!"

"I have always been interested," Mara said, "in the lives of writers, particularly Irish writers. Consider the late Brendan Behan, the well-known Dublin writer. He was born on February 9, 1923, and died on March 20, 1964. Here is a number curiosity that your readers might like. Mara took a card from his pocket and handed it to me. The following was written on the card:

Curiosity About Brendan Behan

By

Mara, the Seafarer

Brendan Behan was born in 1923

Brendan Behan died in 1964

$$19 \times 2 \times 3 + 196 \times 4 = 2^4 + 1^4 + 4^4 + 5^4$$

Brendan Behan lived for exactly 2,145 weeks

"That curiosity is new to me," I said.

"Give it to your readers," Mara said. "They may enjoy it. One Irish writer I greatly admire is George Bernard Shaw. He once said, 'Some men see things as they are and ask why. Others dream things that never were and ask why not.'"

"I know of that quotation," I said. "When President John F. Kennedy visited Ireland in 1963, he referred to that quotation."

"I believe he did," Mara said. "Shaw was born in 1856 and died in 1950. The sum of the squares of the digits in his year of birth is 126. The 126th prime number is 701. The reverse of that number is 107. That number happens to equal the sum of the squares of the digits in Shaw's year of death."

"Holy Macintosh!" I said. "How do you come up with these curios?"

"A little joke about GBS goes as follows," Mara said, ignoring my interjection. "Shaw was once propositioned by a very attractive young woman who was not—let us say charitably— academically gifted. The young woman maintained that she and Shaw should get together and have children. She argued that their offspring would do very well in life because they would have her beauty and Shaw's brains. 'But what would happen,' asked Shaw, who was not considered a handsome man, 'if they inherited your brains and my looks?'"

"That's a good one," I said. "You must have done a lot of research on Shaw."

"I did a little research," Mara admitted, "on GBS. While I was at it, I also read a little of the life of Laurence Sterne."

"I never heard of him," I said.

"He is one of many people," Mara said, "whom you have never heard about! Let me briefly fill you in. Laurence Sterne was a novelist and an Anglican clergyman. He was born in Clonmel, County Tipperary, here in Ireland, on Tuesday, November 24, 1713 (Old Style Calendar), and died on Friday, March 18, 1768 (New Style Calendar). I like many of his quotations. Sterne once said, 'Respect for ourselves guides our morals; respect for others guides our manners.' I just wish that more people in the world would live according to that principle."

"I would certainly agree with Sterne's quotation," I said as I scribbled it in my notebook.

"Glad to hear that," Mara said. "Here is a curious equation I discovered concerning the year of birth and year of death of Laurence Sterne. I have it here on a card. Your readers might find it interesting." Mara reached for his wallet and removed another small, laminated card and handed it to me. The following was written on the card:

Curiosity Concerning Laurence Sterne, the Novelist and Anglican Clergyman

By

Mara, the Seafarer

Year of Birth	*1713*
Year of Death	*1768*

$$17 \times 13 - 1 - 7 - 6 - 8 = 1^2 + 9^2 + 8^2 + 2^2 + 7^2$$

Laurence Sterne lived for exactly 19,827 days.

I read the contents of the card. I found it pretty and impressive. "As we are on the subject of numbers," I said, "I have been told that you like pretty number patterns. Have you any that you would like to share?"

"Sure! I have always liked the fact that the product of four consecutive integers, plus 1, equals a square number. I will give you a few examples." Mara reached for his notebook and pen and wrote down the following on a page:

$$1 \times 2 \times 3 \times 4 + 1 = (2^2 + 1)^2$$

$$2 \times 3 \times 4 \times 5 + 1 = (3^2 + 2)^2$$

$$3 \times 4 \times 5 \times 6 + 1 = (4^2 + 3)^2$$

$$4 \times 5 \times 6 \times 7 + 1 = (5^2 + 4)^2$$

I read the card. "Beautiful!" I said.

"But there is more to this pattern," Mara said. "Let the gap between successive integers equal d. The product of four consecutive integers, plus d^4, is equal to a square number. For instance, when d equals 1, we simply add 1^4, or 1, to the result of the product. When d equals 2, we add 2^4, or 16, to the result of the product. When d equals 3, we add 3^4, or 81, to the result of the product. The result is always a square number. For example, the product of 1, 3, 5, 7 is 105, plus 2^4, gives 121, a square number. The product of 5, 8, 11, 14, plus 3^4, is 6,241. That number is the square of 79."

"That's very interesting," I said.

"Here is a little pattern I discovered when I was drinking some beer with friends during my two months off the ship in a London pub. I am always thinking about co-incidences and numbers, and when I get tired of that, I sometimes ponder the meaning of existence. At the time, when I was having a cool beer, the pattern just occurred to me in a flash. I at once pulled my notebook out of my jacket pocket and wrote the pattern down on a page from it. I still have that page here."

Mara took a tattered page from his pocket. The following was written on the page:

$$2 = 1^3 + 1$$

$$4 + 6 = 2^3 + 2$$

$$8 + 10 + 12 = 3^3 + 3$$

$$14 + 16 + 18 + 20 = 4^3 + 4$$

$$22 + 24 + 26 + 28 + 30 = 5^3 + 5$$

And so on.

I took a note of the pattern, which I had not seen before. Later, I realized that Mara had simply added 1 to each of the terms in a long-ago-discovered series containing the odd numbers, beginning with 1, where each row sums to a cubic number: $1 = 1^3$, $3 + 5 = 2^3$, $9 + 7 + 11 = 3^3$, and so on.

"Do you know of any interesting theorem in number theory?" I asked. "My readers like to read of such theorems from time to time."

"I do. Have you ever heard of the *Euclid-Mullin sequence*?"

"I can't say that I have," I answered. "Should I have?"

"I suppose it all depends on one's interests," Mara quipped. "The Euclid-Mullin sequence begins with the two primes, 2 and 3. The sequence is named after Euclid of Alexandria, who was born around 300 years before the common era, and Albert Alkins Mullin (1933–2017), an American electrical engineer and mathematician who inquired about the sequence in 1963. The third term in the sequence is the least prime factor of 2 times 3 plus 1, which is 7. Since 7 is prime, the third term in the sequence is 7. The fourth term is the smallest factor of 2 times 3 times 7 plus 1, which equals 43. That number is also prime, so the fourth term in the sequence is 43. The fifth term in the sequence is the smallest prime factor of 2 times 3 times 7 times 43 plus 1, which equals 1,807. Its smallest prime factor is 13. Therefore, the fifth term in the sequence is 13, and so on. The first 20 terms of this infinite sequence are 2, 3, 7, 43, 13, 53, 5, 6,221,671, 38,709,183,810,571, 139, 2,801, 11, 17, 5,471, 52,662,739, 23,003, 30,693,651,606,209, 37, 1,741, 1,313,797,957. It can be shown that the sequence does not have any repeated terms."

"I'm with you!" I said.

"Good!" Mara said. "Now for the interesting part. In 1963, Albert A. Mullin at the University of Illinois asked if *every* prime number appears somewhere in the sequence. To this day, no one knows! The vast majority of the terms in the sequence— but not all—are primes. We just do not know if the sequence captures all the primes! We do not even know if we can compute a prime that is in the sequence, except for the algorithm that I have just given. The smallest prime that has not yet been discovered in the sequence is 31. But it may well be in there and may be found if one goes out far enough in the sequence. Only 51 terms of the sequence are known. To figure out the 52nd term requires finding the least prime factor of a 335-digit composite number![1] It would be nice to think that the sequence contains all of the primes. On the other hand, the sequence may not include *all* of the primes. The question is easy to ask, but so far it has defeated the brightest mathematical minds on this planet."

"What's your own view on this?" I asked.

"To be honest," Mara said, "I do not know enough mathematics to make an intelligent guess. I will say this, however. There are a number of possibilities. Perhaps the vast majority of the terms in the sequence are prime. Perhaps there are a finite number of primes that do not appear in the sequence. Or maybe there are an infinite number of primes that are not captured in the sequence. We just do not know! I am inclined to believe that the sequence captures *all* of the primes, but I admit that I have no hard evidence for believing this. It is just a gut feeling I have."

"Extraordinary!" I said. "Thanks for bringing the sequence to my attention. I did not know of it."

"You are welcome," Mara said. "When I was a young lad, I used to spend time investigating number curiosities. Those were the happy, carefree days of yore, when bills came to my house in someone else's name. I discovered a few simple but nice number curiosities back then. For example, consider the sum of the three numbers 987,654,321, 1, and 123,456,789. The answer is 1,111,111,111. When you subtract 123,456,789 from 987,654,321, the same nine digits appear in the answer: 864,197,532."

"I am aware of those results," I said.

"I am strangely proud," Mara said, "of the following insignificant discovery, which I made when I was seven years old: consider the sum of the triangular number, 36, plus the square number 400, plus the cube number, 8. The total of that sum is 444. Using the alphabet code, where A equals 1, B equals 2, and so on, the sum of the letters in the words TRIANGULAR NUMBER, PLUS A SQUARE, PLUS A CUBE also equals 444."

"By heck," I said. "That is a good one. It is new to me. I never read of that before! You are particularly good to retain your interest in recreational mathematics given the fact that you are away from home a lot when you are at sea."

"Speaking of the sea," Mara said, "reminds me of one of the most famous ships in recent history. I speak of the Cunard liner *Queen Elizabeth 2*. It is more commonly known as the QE2."

"The QE2," I said, "is certainly famous. I saw it when it berthed in Cork Harbor in 2008. It was a glorious sight. Large crowds turned up to see this majestic ship."

"I heard about that," Mara said. "I believe the QE2 received a great Irish welcome when it visited Cork Harbor. I checked the statistics of this large liner. The QE2 is 963 feet long and has a breadth of 105 feet and a draft of 32 feet. The QE2 was assigned the yard number 736 when she was under construction. She went on her official first voyage on May 2, 1969. I prepared a card a few weeks back on the statistics of the QE2. I thought you and your readers might be interested in seeing them. I have the card here with me."

Mara reached for his wallet, which was in the inside pocket of his jacket. He took out a laminated printed card. The top right-hand corner of the card had a photograph of the QE2. The card had the following information:

Curiosities Concerning the Cunard Liner, Queen Elizabeth 2

By

Mara, the Seafarer.

Length (in feet) of Queen Elizabeth 2	963
Breadth (in feet) of Queen Elizabeth 2	105
Draft (in feet) of Queen Elizabeth 2	32
When under construction, the QE2 was assigned yard number 736	736
Official first voyage occurred on 122nd day of year.	122
Official first voyage occurred in the 5th month	5
Official first voyage occurred on the 6th day of the week	6
Total equals the year in which the QE2 made her first voyage	1969

I looked at the card. "That information is very interesting!" I said.

"Glad you like it," Mara said. "Here's a curious fact. Using the usual alphabet code where A equals 1, B equals 2, C equals 3, and so on, the sum of the letters in the phrase QUEEN ELIZABETH THE SECOND equals 243. The QE2 went on her official first voyage 243 days before the end of the year."

"Nice one," I said. "Where do you get this information?"

"These curiosities," Mara said, "are original with me. Incidentally, the QE2 belongs to the Cunard Cruise Line. The QE2 commenced its official first voyage on the 122nd day of the year. From the day that the QE2 was launched (on Wednesday, September 20, 1967) to the day that it was officially handed over to its new owners, on Thursday, November 27, 2008, is a period of 15,044 days. The integer part of the square root of 15,044 is 122."

"Amazing! I said. "Did anyone ever tell you that you are great at this sort of thing?"

"Sure! All the time! As the late American footballer Walter Payton (July 25, 1953–November 1, 1999) once famously said, 'When you're good at something, you'll tell everyone. When you're great at something, they'll tell you.' I have always been told that I am great at this! Therefore, I know I must be good at it!"

"I appreciate the fact that you are very modest about your abilities!"

"I am glad you feel that way," Mara said.

"I suppose," I asked, rhetorically, "that seafarers like you are interested only in the lives of those who travel the high seas."

"Not at all," insisted Mara. "I have always taken an active interest in *all* travelers and all explorers. One that comes to mind is the late Sally Ride, the physicist and NASA astronaut born on May 26, 1951. Sadly, Sally died from pancreatic cancer on July 23, 2012. Sally Ride was the youngest American astronaut to be launched into the *final frontier*. She spent 343 days, 7 hours, and 46 minutes in space. Sally Ride lived for 22,339 days. The number of whole days she spent in space is 2 + 2 + 339."

"Brilliant! I was unaware of that information about Sally."

Mara then went ahead to tell me about a famous ship that sank in Lake Superior in 1975. The ship was an American Great Lakes freighter named the *Edmund Fitzgerald*. She commenced her last voyage on Sunday, November 9, 1975. The *Edmund Fitzgerald* sank in a Lake Superior storm the following day, Monday, November 10. All the crew of 29 were lost.

The *Edmund Fitzgerald* was owned by the Northwestern Mutual Life Insurance Company of Milwaukee, Wisconsin. The ship was named by Northwestern after Edmund Fitzgerald, the president and chairman of the board of that company. She was laid down on Wednesday, August 7, 1957, and was launched on Saturday, June 7, 1958. Her launching was dogged with ill fortune. It took three attempts by Elizabeth Fitzgerald, wife of Edmund Fitzgerald, to break a champagne bottle against the bow to christen the ship. The launching was then delayed for 36 minutes as shipyard workers tried to release the keel blocks from under the ship. Finally, when the ship was eventually launched sideways, it crashed into a pier.

Mara said, "The period of time from when the *Edmund Fitzgerald* was first laid down to the day she sank in Lake Superior is a period of 6,669 days. Note the dreaded *number of the beast* in that number! The number 9 indicates that the *Edmund Fitzgerald* commenced her final voyage on the 9th day of the month!"

"Yes," I said, "I am aware of 666. You don't believe all that nonsense about it being the number of the beast, do you?"

"Well, I am not sure. I suppose you know that seafarers are a superstitious lot. Perhaps that explains my uneasiness with the number 666. In any event, the *Edmund Fitzgerald* had a lifetime of 6,669 days. It sank beneath the waves in 1975. Double that number, and one obtains 3,950, whose digits sum to 17. The 17th prime is 59, which equals 13 plus 19 plus 27. The product of those three numbers is 6,669."

"Nice one," I said.

"Gordon Lightfoot," Mara said, "the Canadian singer and songwriter, composed a song titled 'The Wreck of the Edmund Fitzgerald.' It was released in 1976 and became immensely popular. It was his way of remembering those who perished in the tragedy. There are 29 letters in that song's title, the same number of people who were lost in the accident."

"I find that very interesting," I said.

"Good! I am pleased to report that the story of the *Edmund Fitzgerald* is still remembered. Gordon Lightfoot refers in his song to the belief that the legend of this famous ship lives on."

"Quite so," I said. "As we have mentioned the number 29, have you anything unusual about it?"

"There appears to be," Mara said, "something unusual about all numbers if we can only discover that *something*. As for 29, it equals the following expression, which contains all nine digits from 1 to 9: $(3 + 7) \times 2 + (9 - 4)/5 + 1^6 \times 8$."

"Nice one," I said, as I scribbled the expression in my notebook.

"Before I go any further," Mara said, "let me state that these curiosities that I am giving you are completely original with me, so you will not have come across them before. I just happen to like numbers and coincidences."

"So do I, Mara!"

"Yes, that is why we are here talking! Now where was I? Oh yes! Consider the case of the United Kingdom's late Queen Mother and the palindromes associated with her life and death."

"What palindromes?" I asked.

"Well, let's see," Mara said as he reached into his jacket inside pocket, pulled a page out, and handed it to me. "Here is a list of curiosities concerning the British Royal Family that I put together." The following was written on the page:

The life of the United Kingdom's late Queen Mother and her connection to palindromes

By

Mara, the Seafarer.

Using the usual code, where A equals 1, B equals 2, and so on, the sum of the letters in the phrase BRITISH ROYAL FAMILY equals 222, which is a palindrome.

The Queen Mother died on Easter Saturday, March 30, 2002. The Queen Mother's first name was Elizabeth. Using the code that I mentioned earlier, the sum of the letters of ELIZABETH equals 88. That number is a palindrome. Using the same code, the letters of the title QUEEN MOTHER sum to 141. That number is a palindrome. The Queen Mother's family home was Glamis Castle, which was, incidentally, the legendary home of Macbeth of Shakespearean fame. Using the same code, the sum of the letters in GLAMIS CASTLE equals 121. That number is a palindrome. Using the same alphabet code, the sum of the letters of the phrase EASTER SATURDAY TWO THOUSAND AND TWO (the day of death of the Queen Mother) is 414. That number is a palindrome. Using the same code, the sum of the letters in the phrase DEATH OF A ROYAL equals 131. That number is also a palindrome.

Curiously, there are 11 letters (11 is a palindrome) in the two words QUEEN MOTHER. During her tenure as Queen Consort, the Queen Mother was known by the title HER ROYAL HIGHNESS, THE DUCHESS OF YORK. Using the earlier code, the sum of the letters in that title equals 393. The number 393 is a palindrome. The Queen Mother died on March 30, 2002. The date March 30 is usually written in the United Kingdom as 30/3. The number 303 is a palindrome. The Queen Mother died in 2002. The number 2,002 is a palindrome. The Queen Mother died on the 454th day of the new millennium. The number 454 is a palindrome. The Queen Mother was laid to rest on April 9, which was the 99th day of that year. The number 99 is a palindrome. The Queen Mother was laid to rest on the 464th day of the new millennium. The number 464 is a palindrome. The Queen Mother was 101 years old when she died. The number 101 is also a palindrome.

The House of Windsor is the *current royal house of the Commonwealth realms*. Using the earlier alphabet code, the sum of the letters in the words HOUSE OF WINDSOR is 191.

I read the page. "That is most interesting," I said. "I have never read or even heard about these curiosities before."

"I can understand why that is the case," Mara said. "I never published them before! They are completely original with me!"

"You have a very unique mind, Mara."

"Thank you!" Mara said. "I will take that as a compliment."

"It was meant as a compliment," I said. "All these palindromes remind me of a nice palindrome that others have discovered over the years. Using the usual alphabet code, the sum of the letters in the phrase PALINDROMES ARE FUN equals 191, which is a palindrome prime."

"That is wonderful!" Mara said. I could see that he was genuinely delighted with my little contribution.

"You have a great intellect," I said. "It is amazing how you discovered those curiosities concerning the late Queen Mother."

"I suppose I have," Mara said modestly. "I like collecting unusual facts. I can retain this type of information. When I was younger, my memory was brilliant. But even now, it is not too bad. But sometimes I get disorientated, and my memory diminishes as a result. When I was in my early teens, I would sometimes go hiking in the mountains. I found the fresh air and peace and quiet there so delightful. Of course, I always made sure I would bring my watch with me in case I got lost."

"Your watch!" I said. "How would that help you if you got lost?"

Mara then conveyed to me an interesting piece of information. I took careful notes of what he was telling me. He pointed out to me how an analog watch may be used as a compass in the northern hemisphere. To do so, hold the watch flat in the palm of one's hand so that the hour hand is pointing toward the sun. If the time is forenoon, then the point halfway between 12:00 and the hour hand points toward the south. For example, if the time is 8:00 a.m., point the 8 in the direction of the sun. The 10 on the watch face will now be pointing south.

If the time is in the afternoon, bisect the backward angle between the hour hand and the 12:00 mark. For example, if the time is 4:00 p.m., then the 2 on the watch face is pointing south.

If it is midday, point the 12 toward the sun. The 12 is then pointing south. If daylight savings time is in operation, the same procedure applies, except use 11:00 instead of 12:00.

If one is in the southern hemisphere, a similar procedure applies. Hold the watch flat in one's hand so that the 12:00 mark points in the direction of the sun. The point halfway between the hour hand and the 12:00 mark points north. In the forenoon, at 8:00 a.m., for example, the 10 on the watch face is pointing north. In the afternoon, at 4:00 p.m., for example, the 2 on the watch face points north. The procedure in both hemispheres works best the farther one is from the equator. The directions are not exactly correct, but, Mara said, they are particularly good approximations.

"Life and time," I said after listening to this most unusual character, "are such strange things."

Mara responded philosophically: "Our life, as Thomas Carlyle (1795–1881) once said, 'is a little gleam of time between two eternities.'"

"You got that in one," I said.

I looked at my watch. "I will have to go soon," I said. "Before I leave, have you any little number puzzle for my readers?"

"Let me see. Ah yes, I have one. If we square the expression $(x + y)$, our answer is $(x^2 + 2xy + y^2)$. Thus, it is natural to think that all perfect square numbers are of this form. The question that I pose is this: can perfect square numbers be of the form $x^2 + xy + y^2$?"

"I don't know," I said. "I'll give that to my readers."

"Here is a little puzzle," Mara said, "with a nautical flavor. I think your readers may like it. What goes down one side of a ship and goes in the opposite direction on the other side of the ship?"

"I'll give that to my readers also," I said.

With that, I put my pen and notepad back in my pocket. I shook hands with this most unusual and good-humored mariner, promised to stay in touch, and bade him a long farewell.

Before I knew it, I had left the comfortable hotel and was walking briskly through the busy streets of central Dublin.

Solutions

1. Mara asked if a perfect square number can be of the form $x^2 + xy + y^2$.

   The surprising answer is they can! If, for example, we make x equal 5 and y equal 3, we find that $5^2 + 3 \times 5 + 3^3 = 49 = 7^2$. There are an infinite number of solutions.

   From the question, let $a^2 + b^2 + ab = (a - mb)^2$.

   Thus, $a^2 + b^2 + ab = a^2 - 2amb + m^2b^2$.

   Or $b^2 + ab = -2amb + m^2b^2$.

   This equation can be rearranged as

   $$\frac{m^2 - 1}{a} = \frac{2m + 1}{b}$$

   Thus,

   $$b = \frac{a(2m + 1)}{m^2 - 1}$$

   Let m equal any integer greater than 1 and choose a to make b rational in the following equation:

   $$b = \frac{a(2m + 1)}{m^2 - 1}$$

   Thus, if we make m equal 2, we find that to make the equation equal a rational number, we make a equal to 3. We then have

   $$b = \frac{15}{3}$$

   The equation gives b a value of 5. Thus, our two smallest numbers for a and b are 3 and 5. Our equation then is $5^2 + 3 \times 5 + 3^3 = 49 = 7^2$.

   Generally, make $a = m^2 - 1$ and $b = 2m + 1$.

   Thus, if we make $m = 3$, $a = 8$, and $b = 7$, we obtain the second-smallest solution: $8^2 + 8 \times 7 + 7^2 = 169 = 13^2$.

   Thus, if we make $m = 4$, $a = 15$, and $b = 9$, the third-smallest solution is $15^2 + 15 \times 9 + 9^2 = 441 = 21^2$ and so on.

2. Mara asked, "What goes down one side of a ship and goes in the opposite direction on the other side of the ship?"

   Answer: the name of the ship!

CHAPTER 18

Similarities and Curiosities between the RMS *Titanic* and Eastern Air Lines Flight 401

At 11:40 PM on April 14, 1912, the *Titanic* struck an iceberg and sank in the early hours of the following morning.

At 11:40 PM on December 29, 1972, Eastern Air Lines Flight 401 crashed into the Florida Everglades.

Both the *Titanic* and the Eastern Air Lines Flight 401 disasters occurred in years that leave a remainder of 4 when divided by 12.

Both disasters occurred on a moonless night.[1]

Both disasters occurred shortly after major Christian festivals.

Both disasters occurred at or very close to 11:42 p.m.[2]

Eastern Air Lines Flight 401 crashed into the Florida Everglades.[3] The Everglades is a form of marsh.

The captain of the *Titanic* was Edward John Smith. His mother's maiden name was Marsh.[4]

The captain of Flight 401 was Robert Albin Loft.

Both the captain of the *Titanic* and the captain of Eastern Air Lines Flight 401 had surnames that are nouns in the English language.

The band on the *Titanic* played right up to the end in an effort to keep passengers calm.[5]

Survivors of Eastern Air Lines Flight 401 sang Christmas hymns in the Everglades in an effort to keep each other calm.[6]

The *Titanic* struck an iceberg on 4/14.

Eastern Air Lines Flight 401 crashed on 12/29. Note that $414 + 414 + 401 = 1,229$.

There were 29 boilers on board the *Titanic*.

Flight 401 crashed on December 29.

Eastern Air Lines Flight 401 was a relatively new airplane that had been in operation for just four months before crashing.

The *Titanic* sank on her maiden voyage.

The *Titanic* was built in the Harland and Wolff Shipyard in Belfast. The yard number of the *Titanic* was 401.[7]

CHAPTER 19

Charlie Chance, the Mathematical Barman

Many years ago, I was over in Manchester, in the United Kingdom, when I met a rather strange character. He told me his name, but he told me he would prefer if I referred to him by the fictional name he had adopted: Charlie Chance. He had adopted this name because of his love of spotting strange coincidences. In any event, I spent two hours or so in Charlie's very pleasant company on that day.

I had not seen or heard of Charlie since that chance meeting in the relatively distant past. About three months ago, Charlie was in contact by phone. How he obtained my personal number is a mystery, the solution of which is probably better left for another day. In any event, Charlie told me he was back in England, Manchester to be more precise, having been abroad for several years. He told me he had spent the previous few years working as a barman in Spain, where he had acquired the name the "Mathematical Barman." Charlie gave me his phone number and asked me to call on him if I was ever in Manchester.

As things happened, I had some confidential business to attend to in Manchester about one week ago, so I took Charlie up on his offer.

Last Friday morning, I got on board a flight from Dublin that took me on the short journey across the Irish Sea into Manchester Airport.

I made my way to a city-center hotel, where I had arranged to meet Charlie.

I met Charlie as I waited in the foyer of the hotel. I instantly recognized him as he walked in. He was still as tall and as lean as ever, about six feet tall and weighing about 160 pounds. He had a short, neat haircut that still showed off his jet-black hair. I arose from my seat and introduced myself.

"Charlie, how are you?" I asked.

"Owen, it is lovely to see you again, mate. I am great. How are you?"

"I am fine also."

"I hear you are a writer these days."

"Well, I try," I said.

"I hear you write about recreational mathematics if the rumors that I hear are true. Of course, I knew you were always interested in that subject. I am only a moderate mathematician, but I am, as you know, fascinated by strange coincidences. I know you share my interest."

"I certainly do, Charlie," I said. "Listen, why don't we grab a couple of beers and have a little chat. I might be able to use some of your coincidences in my forthcoming book. If I do, I will, of course, credit the coincidences to you."

"That would be nice," Charlie said. "But only use my pseudonym, 'Charlie Chance.' Otherwise, I will be inundated with letters about this and that from all sorts of people. I am a busy man. I could not cope with that type of fan mail."

"Nor I," I said as we wandered into the lounge bar and ordered two cool beers. We sat over by a window. Our beers were brought to us very promptly.

I took out a notebook and pen.

"I see you have come prepared," Charlie said. "I hope I will not disappoint you."

"From what I know of you, Charlie," I said, "there is little *chance* of that happening."

"Thank you, my friend," he said. "I appreciate your kind words. Incidentally, I am sure that there is no pun intended in what you just said."

"None whatsoever!" I replied quickly but dishonestly. "Now let me get down to business. Why did you come to live in Manchester?"

"I have always liked Manchester. I am not sure why but maybe because it has an interesting history. Manchester was the world's first industrial city. The movement to secure voting rights for women began in Manchester. Ms. Emmeline Parkhurst and her daughter Christabel founded the Women's Suffrage League there. They later founded the Women's Social and Political Union. They succeeded in bringing in political reform. By 1918, women over 30 years of age had secured the right to vote. By 1928, women of 21 years of age were entitled to vote."

"That's interesting information," I said. I scribbled notes of what he had said in my notebook.

"Are you still interested in soccer?" I asked.

"You can bet your bottom dollar I am," he said. "I know you are too! Anyone who knows anything about Manchester, Owen, knows that Manchester is home to probably the most famous soccer club in the world? I speak, of course, of Manchester United. They are more affectionately known to their supporters as MAN UNITED. They were the first English football club to be crowned Champions of Europe when they won—what was then known as the European Cup—in 1968."

"Yes, Charlie, I am aware of the successful record of Manchester United. Of course, there is also a second famous and phenomenally successful club in this town: Manchester City. They are old rivals to United. But I always knew that you are a United supporter. Are you going to tell me that you have some interesting curiosities on that great soccer club that I can use in my forthcoming book?"

"I certainly am, Owen." I waited in joyful anticipation. I know from experience that when Charlie gives you a coincidence, it is usually a particularly good one.

"On Thursday, February 6, 1958, on their way home from a soccer match in Belgrade, Yugoslavia, the plane carrying the Manchester United team and officials stopped at Munich Airport to refuel. Munich was then part of West Germany. The plane carrying the Manchester United team crashed on the icy runway in the middle of a snowstorm. There were 44 people, including players, club officials, and journalists, on board the plane. Most of that marvelous soccer team were killed."

"Yes," I said. "I am aware of that tragedy! The fantastic Manchester United team was nearly wiped out."

"Yes indeed," Charlie said. "It was a devastating blow to the team that was known as the 'Busby Babes.' They had been given that name because they were a relatively young team, managed by the legendary Matt Busby (1909–1994). Anyway, here's a little curiosity that your readers might enjoy concerning the first letters of the word MUNICH." Charlie wrote the following down on a page from his notebook and handed it to me:

> **M**anchester
> **U**nited
> **N**ever
> **I**ntended
> **C**rashing
> **H**ere.
> by
> Charlie Chance

"That's a nice little word curiosity, Charlie," I said. "Have you any more curiosities?"

"Of course, I have more, Owen."

"I would love to hear about them, Charlie."

"No problem, my friend," said Charlie, "As you may recall, it was British European Airways Flight 609 that the Manchester United team was aboard on the fateful day of the crash. There were 44 people on board the plane. It crashed on its third attempt at a takeoff on an icy runway. Of those 44, 21 were killed on board the plane in the crash. Two more people later died in hospital, bringing to 23 the total number who lost their lives because of the accident. The crash occurred in 1958. It is curious that $1 + 9 + 5 + 8$ equals 23, the number of fatalities in the crash. Four of the team's survivors were the popular players Harry Gregg, Bobby Charlton, Bill Foulkes, and Jackie Blanchflower. The jerseys they regularly wore playing for Manchester United were usually numbered 1, 9, 5, and 8, respectively."

"My goodness," I said. "Those are the same digits that are contained in the year of the crash!"

"Now you have it," Charlie said. "As I mentioned, there were 44 people on board the plane. The 44th prime is 193. Consider the digits of that number. It is curious that $1,931 + 9 \times 3$ equals 1958, the year of the crash. The airplane accident occurred on the 37th day of the year. Strangely, the number of survivors of the crash equals 3 times 7. The flight number of the plane carrying the Manchester United team was 609. That is a strobogrammatic number. In other words, it reads the same upside down. The 609th prime number is 4,483. Partition that number into *three* parts as follows: 44, 8, and 3. There were a total of 44 persons on board British European Airways Flight 609 when it crashed in Munich. It is curious that 8 of the Manchester United team died because of the plane accident, and 3 of the Manchester United Club Staff were also killed."

"Holy Moses!" I said.

"I mentioned," Charlie said, "that the number 609 is strobogrammatic. Well, I should also point out that the number of primes less than 609 is 111. That number is also strobogrammatic."

"Amazing!"

"If one reverses the number 1,958," Charlie said, "one obtains 8,591. The sum of the proper divisors of 8,591 is 985. The sum of the proper divisors of 985 is 203, and the sum of the proper divisors of 203 is 37. The Munich plane crash occurred on the 37th day of the year!"

"Astonishing!" I said. "That result is not at all obvious! How did you figure that out?"

"I can't really explain it," Charlie said, "I went to bed one night wondering if there was a connection between 1,958 and 37. The solution came to me in a dream! Here is a little curiosity that I think your readers will like. I never published it before." Charlie took a card from his pocket, looked at it for a moment, and then handed it to me. I looked at the card, on which the following was written:

A Curiosity Involving Manchester United Football Team
by
Charlie Chance

Using the following code where A = 1, B = 2, C = 3, and so on, sum the values of all the letters in the following two sentences:

Manchester United crashed in Munich Airport on their third attempt at taking off on a Thursday, February Sixth, Nineteen Fifty-Eight. Eight of that great team were tragically killed that fateful Thursday.
The sum of all the letters in those two sentences equals 1,958!

"Before giving you the next curiosity," Charlie said, "I will point out to you that Manchester United are often known by their nickname the 'Red Devils.' Here is a little curiosity that I spotted concerning Manchester United in more recent years. You may be aware that by February 4, 2011, the Red Devils had gone 29 Premier League games in a row in the football season without defeat. That means the Manchester United side of 2011 had equaled their own club record of 29 Premier League games in a row without defeat, which was achieved by the Red Devils team of 1999."

"I see," I said. "What's the curiosity?"

"I'm coming to it," Charlie said. "Have patience, mate! On Saturday, February 5, 2011, Manchester United played the Wolverhampton Wanderers. The number 29 was very much in the news in relation to this game. The sports columnists in the newspapers were asking the question that was on the lips of every Manchester United fan: could the Red Devils surpass their club record of 29 Premier League games in a row without defeat? All Manchester United fans were hoping their team would win, thereby achieving a new club record of 30 Premier League games without defeat. However, the Red Devils lost the game by 2 goals to 1. The winning goal for Wolverhampton was scored by Irishman Kevin Doyle. Curiously, he was wearing the number 29 jersey that day, and he scored the winning goal in the 29 + 2 + 9 minute of the game!"

"Holy jeepers," I said. "That *is* curious! You mentioned, incidentally, that the game was played on 02/05. Of course, the sum of the squares of 02 and 05 equals 29."

"That is clever of you to spot that," Charlie said.

"Here is a fantastic curiosity," I said, "given to me by an old school friend in Cobh named Michael O'Leary. He is an old friend of my twin brother, Michael, also. We all grew up together. The curiosity concerns the professional French soccer player Thierry Henry. (His first name is pronounced 'Thi-erry,' and his surname is pronounced 'On-ree.') On January 9, 2012, Thierry Henry was the 12th man (or substitute) for the London soccer team, Arsenal, in their game against Leeds United. Appropriately, Henry was wearing the number 12 shirt when he came on as a substitute in the match. The Arsenal manager decided to introduce Henry into the game in the 68th minute, with the score nil all, and just 22 minutes to go to full time. In the 78th minute, with just 12 minutes to go, Thierry Henry scored a brilliant goal that won the game for Arsenal. Here's the curiosity that my friend gave me: Henry was wearing the number 12 shirt when he scored the winning goal with just 12 minutes of the match remaining; it was Henry's 12th Arsenal goal in 12 appearances against Leeds United. The goal was scored in the 12th year of the century!"

"Yes," Charlie said, "I read about that amazing coincidence in the newspapers at the time. It was strange! In fact, it was amazing! It was a magical night for Thierry Henry, for the Arsenal team, and for their supporters. It was a night in which Arsenal supporters who were there at the Emirates Stadium that day to see the magic, will, in years to come—when they hear talk of that famous night in January 2012—say 'I was there!' Incidentally, Henry scored the goal from 12 yards from the Leeds United goal line in the 78th minute. No one appears to have noticed that the sum of the numbers from 1 through 12 equals 78. Also, Henry was sporting a beard on the night he scored that glorious goal. Apparently, he had not shaved for 12 months."

"That *is* truly weird!" I said. "It must go down as not only one of the most famous and strange coincidences in soccer but also one of the strangest coincidences ever recorded. I believe Henry has said that this goal is his favorite goal of his career."

"That is correct," Charlie said. "It is easy to see why! That particular game was Thierry Henry's 12th Arsenal match against Leeds United in his soccer career. The initial letter of Leeds United is, of course, the 12th letter of the alphabet. That particular goal was Henry's 12th Arsenal goal against Leeds United in Henry's soccer career. He was 34 years old at the time. The product of 3 and 4 is 12. The two phrases JANUARY NINTH and TWENTY TWELVE each contains 12 letters. The name Thierry Henry has 12 letters. His initials are *TH*. Those are the 20th and 8th letters of the alphabet. The difference between those two numbers is 12."

"Incredible! I never spotted those curios," I said. "It is an amazing set of coincidences involving Henry and the number 12."

"It sure is," Charlie said. "Here's something else that you probably are not aware of. Thierry Henry scored his famous magical goal on January 9, 2012. That date may be written as 1/09/2012. The number 1,092,012 is evenly divisible by 12."

"That's brilliant," I said. "By the way, in England, the date of January 9, 2012, would be written as 9/01/2012. Is 9,012,012 evenly divisible by 12?"

"It certainly is!" Charlie said.

"Astonishing!"

"The number 2012," Charlie said, "will probably hold a special significance in Henry's life as a consequence of that magical goal he scored. The number 12, I believe, may very well be described as Henry's lucky number. Here's a curiosity. It involves the primes and the number 37, which is the 12th prime. In any event, let me tell you that Henry was born on August 17, 1977, which may be written as 17/8/77 (as they write dates in England). Have a look at this table that I devised." Charlie took a small, laminated page from his pocket and handed it to me. The following was written on the page:

A Curiosity Concerning Thierry Henry
by
Charlie Chance

Year in which Thierry Henry was born	1977
Thierry Henry's lucky number is 12; the 12th prime is 37	37
Thierry Henry was 34 years old when he scored his famous goal against Leeds United	34
Total equals	2048

Thierry Henry was born on the 17/8/77. There are 2048 primes less than 17,877.

I looked at the page. Before I could say a word, Charlie had spoken again. "Here is another curiosity concerning Thierry Henry. Your readers might like it for their files." Charlie reached into his jacket pocket. He again took out a small, laminated page on which the following was written and handed it to me:

A Second Curiosity Concerning Thierry Henry
by
Charlie Chance

Year in which Thierry Henry was born	1977
Thierry Henry's lucky number is 12; the 12th prime is 37	37
Thierry Henry scored his famous goal against Leeds United in 2012	2012
Total	4026

Total equals the day-number of century when Thierry Henry scored his famous goal against Leeds United.

I looked at the page.

"That's brilliant!" I said. "I notice, by the way, that the digits of 4026 sum to 12."

"That's sharp of you, Owen," Charlie said. "You're getting good at spotting the unusual. Keep it up! You might get famous someday."

"I doubt that," I said.

"Who knows?" said Charlie. "No one knows what the future holds! Now for something different. I will give you some original curiosities concerning the famous Swedish pop group ABBA."

"Please do!" I said. "I love listening to ABBA. I have no problem in admitting that ABBA is one of my favorite groups of all time."

"Owen, my friend, for countless people around the world, ABBA is the most favorite group of all time."

"Yes indeed," I said. "They were brilliant! All their songs were massive hits! I have been a fan of ABBA all my adult life."

"They were fantastic," Charlie said. "Let me see. Oh yes! Obviously, the name ABBA forms a palindrome. But there are many palindromes associated with ABBA that even the members of ABBA are probably unaware of. They won the Eurovision Song Contest in Brighton, England, on Saturday, April 6, 1974. That success catapulted ABBA to world-wide stardom. The next day, ABBA began life as a super and world-famous pop group. The first full day that ABBA became stars was 4/7/74 (U.S. style of dating). Note that the date is a palindrome! Also, one of their many super hits was called "S.O.S.," which is also a palindrome! Using the usual alphabet code where A equals 1, B equals 2, C equals 3, and so on, the sum of the letters in the phrase A SWEDISH POP GROUP is 212. Note that 212 is also a palindrome. Using the same code, the sum of the letters in the phrase EUROVISION SONG CONTEST IN BRIGHTON equals 414, which is yet another palindrome! The Saturday in which that particular Eurovision Song Contest was held in Brighton in 1974 was the 171st Saturday of the decade. Note that 171 is a palindrome. Note also that the product of the digits in the number 1,974 is 252, which is another palindrome! Agnetha Fältskog, the blonde female singer with ABBA, is also known as ANNA. That name is also a palindrome."

I was scribbling all this fascinating information down as quickly as I could. "This is marvelous information, Charlie," I said.

"I'm not finished! The world-famous band ABBA was named after the initials of the first names of its four members, Agnetha Fältskog, Björn Ulvaeus, Benny Andersson, and Anni-Frid Lyngstad. Their initials are A, F; B, U; B, A; A, L. Using the earlier code, the sum of those letters is 46. ABBA won the Eurovision Song Contest on 4/6, as dates are written in the United States."

"Suffering snowballs!" I said. "That is amazing!" I could hardly believe all this brilliant information, but Charlie assured me that it is all true.

"Finally," Charlie said, "the blonde female singer in ABBA, Agnetha Fältskog, was born on April 5, 1950. That date is written in the United States as 4/5/1950. When ABBA won the Eurovision Song Contest on April 6, 1974, Agnetha was $4 + 5 + 1 + 9 + 5 + 0$ years old!"

"Holy mackerel, those are the digits of her date of birth."

"Precisely," Charlie said. "I hope you liked those curiosities concerning ABBA."

"They are outstanding, Charlie! I will use them in my forthcoming book and will of course credit you with them. I would never be able to discover these."

"I am sure you could if you put your mind to it! In any event, let's move on. You obviously know of Frank Sinatra. He was a world-famous singer. He was known as Ol' Blue Eyes."

"Yes," I said. "Please give me some credit. Of course, I remember him."

"Ol' Blue Eyes," Charlie said, "was born on Sunday, December 12, 1915, and died on Thursday, May 14, 1998. Francis Albert Sinatra was born on the 12th day of the

12th month. The name FRANK SINATRA contains 12 letters. He died on the 134th day of the year. The product of the digits of that number equals 12. His wife stated in her book *Lady Blue Eyes: My Life with Frank Sinatra* that Francis Albert showered 12 times each day. Frank lived for 30,104 days. The product of the nonzero digits of that number equals 12. Ol' Blue Eyes was born on the 346th day of the year. Frank Sinatra died on the 35,563rd day of the century. Of course, note that 346 equals 355 minus 6 minus 3."

"By the beard of the Wise Prophet, you are right!" I said.

"I will give you something different now. I am not sure if you know that Tecumseh was a leader of the Shawnee tribe back in the days of the Wild West."

"I can't say I knew of that," I said.

"Well, that is not a problem. I think you will still enjoy the curiosities I am going to give you. Tecumseh collaborated with other Indian leaders to oppose white settlers expanding their presence into the Wild West. At the 1811 battle at Tippecanoe, U.S. troops led by future president William Henry Harrison defeated Indian warriors, led by Tecumseh's brother, Tenskwatawa. The following year, Tecumseh was killed fighting alongside the British in the Battle of Thames in the War of 1812."

"I am enjoying this history lesson," I said.

"Good," Charlie said. "You will probably enjoy what is to come. It was said that after Tecumseh's death, his brother, Tenskwatawa, predicted that if Harrison ever became the U.S. president, he would die in office and that every U.S. president thereafter elected in a year ending in zero would also die in office. It was never proved that Tenskwatawa ever said this, but his prediction, for want of a better word, became known as the 'Curse of Tecumseh.'"

"Did the so-called curse work?" I asked.

"Well, I will let you make up your mind on that. William Henry Harrison was elected U.S. president in 1840. He was 68 years old when he gave his inaugural speech on March 3, 1841. The day was bitterly cold, but Harrison spoke for about one hour and 40 minutes. He was not dressed suitably for the weather, wearing no coat or hat. Within the next couple of weeks, President Harrison caught a cold, which then developed into pneumonia. He died on April 4 after being in office for one month and one day."

"Gosh!"

"As you know, the election for U.S. president is held every four years. Thus, there was no election in 1850. But there was in 1860. That year, Abraham Lincoln was chosen. He was assassinated on April 14, 1865."

"A terrible loss to not only America but to the world!" I said.

"I completely agree. In 1880, James Garfield was elected president. He was assassinated in 1881. William McKinley was chosen in 1900. He was also assassinated. Warren G. Harding was picked in 1920. He died in office. In 1940, Franklin D. Roosevelt was chosen. He also died in office."

"I think I can see a pattern emerging," I said.

"Yes, you probably can," Charlie said. "John F. Kennedy was chosen in 1960. He was assassinated in November 1963. Finally, in 1980, it appears that the so-called curse was broken. Ronald Reagan was elected president that year. He did not die in office, but he did have a close shave with death. He survived an assassination attempt in March 1981."

"Amazing information!" I said. "But you do not believe in curses, do you?"

"Not at all," Charlie said. "I put the whole thing down to coincidence."

"So do I! Any other coincidences that I should know about?"

"Scores! But before I come to the coincidences, please let me preface my remarks by mentioning a terrible crime against humanity. I speak of the dreadful London bombings on July 7, 2005. Three bombs exploded on trains in Underground railway stations, and one bomb exploded on the upper deck of the number 30 double-decker bus in Tavistock Square. Those terror attacks left 52 innocent people dead and hundreds injured. The four suicide bombers were also killed. Those bombings are now usually referred to in London as the '7/7 bombings.'"

"I recall the bombings," I said. "They were cowardly and dreadful."

"Yes, they were," Charlie said. "My sympathies go the families of the dead and wounded. It was a cowardly attack on innocent people."

"It certainly was," I said.

"Incredibly," Charlie said, "a private crisis management advice company named Visor Consultants was actually running an exercise for another private company of more than 1,000 people in London at 9:30 a.m. on the morning of 7/7 in 2005. The terror bomb drill was based on a scenario where simultaneous bombs explode on trains in Underground railway stations in London. Astonishingly, the railway stations chosen for the mock terror drill were precisely those railway stations where bombs actually went off on Underground trains on the morning of 7/7! As part of the exercise, a room full of crisis managers had met for the first time at about 9:30 a.m. on the morning of 7/7. News then came through that trains in the actual Underground railway stations in London that had been chosen for the mock terror drill had actually been bombed."

"Gosh," I said. "That was an amazing coincidence!"

"Yes, it certainly was," Charlie said. "Former Scotland Yard Officer Peter Power was the managing director of Visor Consultants at the time. On the morning of 7/7, he was one of those involved in organizing the bomb drill exercise. He gave a radio interview to the British Broadcasting Corporation (BBC Radio 5 Live) on the evening of Thursday, July 7, 2005. During the course of the interview, Power recalled the incredible coincidence for the listeners of the radio program. Power went on to state that the hairs on the back of his neck were still standing up as a result of the coincidence."

"Amazing!" I said.

"There was another series of coincidences," Charlie said, "on that dreadful day. A fourth bomb exploded on the upper deck of the number 30 double-decker bus in Tavistock Square. The bus in question had been forced to take a detour because of the drama surrounding the bomb attacks on the Underground railway system. Incredibly, the bus exploded in front of the offices of the British Medical Association (BMA) in Tavistock Square. By a strange coincidence, there was a medical conference being held at the offices of the BMA that very morning! Consequently, dozens of doctors were on hand to help the wounded in the bus explosion. One of the general practitioners at the conference was Dr. Peter Holden. He is trained as one of the few major incident commanders in the United Kingdom. So, crucially, he was on hand to help also."

"That was another truly amazing coincidence!" I said.

"Yes, it was!" Charlie said. "The 7/7 bombings killed 52 innocent people. The digits of 52 appear in 2005, the year of the attack. The number 52 plus its reversal equals 77, whose digits appear in the date of the attack. The number 77 plus its reversal equals 154. The sum of the proper divisors of 154 is 134, which equals $1^2 + 6^2 + 4^2 + 9^2$. The 7/7 bombings happened on the 1,649th day of the century."

"Amazing!" I said. "I hope you don't mind me saying that it takes a weird sort of mind to come up with these figures."

"I am used to comments like that. You must remember that I worked as a barman for a number of years in Spain. I met all kinds of people—some good and, well, let's say, some not so good. I guess it is the same the world over. I remember I told a number of Spanish customers that the Spanish name for 'five' is *cinco* and that cinco has five letters. Thus, cinco is a very *honest* number. Some liked what I told them. Others could not care less. I guess it takes all sorts to make a world."

"I know what you mean," I said. "I get similar reactions! You probably know that the only *honest* word in English is *four* because *four* has four letters. The only honest word in Irish is *trí*. It means 'three,' and it contains three letters."

"Nice one," Charlie said. "I hope your readers find the details I give interesting."

"I'm sure they will," I said. "I could listen to you all day, Charlie. However, I'm afraid I will have to leave shortly."

"Before you go, Owen, here's one final curiosity. As a barman, I hear a lot of stories: some true and some untrue. I am always interested in hearing or reading true stories of the sea and of famous ships. Consider the *Titanic*, one of the most famous ships that ever sailed the seas. The *Titanic* hit an iceberg on April 14, 1912, at 11:40 p.m., ship's time. Two hours and 40 minutes later, at 2:20 a.m. on April 15, 1912, the famous ship sank to the bottom of the North Atlantic Ocean. According to *The Mirror* newspaper online website, updated October 20, 2016, the number of people who perished in the sinking of the *Titanic* was 1,521.[1] This number equals 39^2. If that figure is correct, one can say that when the Titanic sank on 4/15/1912, the number of people who perished was coincidentally $(4 + 15 + 19 + 1)^2$.

"Gosh! The number of people who died on the *Titanic* can be expressed in a form that gives the date of the sinking of the famous ship! That is truly amazing!"

"Now you have it," Charlie said. "Strange things happen! I guess it is a funny old world!"

"By heck," I said, "it most certainly is! Thank you, Charlie, for giving me the most useful and extraordinary information. I am sure my readers will enjoy reading about these curiosities."

"I hope they will," Charlie said. "I am working on a little project at the moment. I will send you an e-mail in the very near future concerning a set of curiosities I have discovered. You may be able to use them in your forthcoming book."

"Thank you, Charlie," I said. "I look forward to reading those when they arrive."

We exchanged each other's contact information.

Then we shook hands, promised to stay in touch, and said our good-byes.

As I left the hotel, I felt very pleased indeed to have met up with Charlie after all these years. It was a *chance* encounter that I shall always remember.

CHAPTER 20

Coincidences Surrounding King Richard III of England

Two weeks after meeting the fictional Charlie Chance, an e-mail arrived from him, as promised.

The e-mail contained a curious set of number coincidences that occurred in the search, in August 2012, for the remains of Richard III, the last British king to die in battle and the only British king to do so since Harold II was killed at the Battle of Hastings in 1066. Charlie said that the coincidences were entirely original with him and had not been published elsewhere. Richard III was killed in the Battle of Bosworth Field, England, on Monday, August 22, 1485. The search for his remains was narrowed down to a parking lot in the center of the city of Leicester, England. The digging for his remains commenced on Saturday, August 25, 2012.

It turned out that Richard III was buried directly beneath a white letter *R* (presumably representing the word *reserved*, for parking) that had been painted on the tarmac of the parking lot. Philippa Langley, a prominent member of the Richard III Society, which helped fund the search for the dead king, said that she felt she was walking on Richard III's grave as she walked near the painted letter *R* in the parking lot. She felt instinctively that that was where the dead king lay. Incredibly, Philippa Langley was proved right!

On that very first day of digging, they found a leg, virtually under the painted letter *R*, that turned out to be part of Richard III's body. The rest of the king's skeleton was found soon after.

The period of time from Richard III's death to the day his remains were found in Leicester on August 25, 2012, is 192,477 days. The skeleton of Richard III was found on the 238th day of the year. Using the usual alphabet code, where A equals 1, B equals 2, C equals 3, and so on, the sum of the letters in the words OLD KING OF ENGLAND IS FOUND equals 238. Curiously, 238 equals $1 - 92 + 47 \times 7$. Using the same alphabet code, the sum of the letters in the words RICHARD THE THIRD equals 153, and the sum of the letters in the words LEICESTER ENGLAND also equals 153. The difference between 238 and 153 is 85. Richard III was killed in the 85th year of the fifteenth century.

CHAPTER 21

Nine Problems

The following are nine problems that are not too difficult and that I trust readers will enjoy tackling:

1. I have written down a five-digit number. If I place 1 at the end of the number, the resulting number is three times larger than if I put a 1 at the front of the five-digit number. What is that five-digit number?
2. Two missiles are flying toward each other. One missile is traveling at 21,000 miles per hour. The second missile is traveling at 9,000 miles per hour. Soon after the two missiles were fired, they were 2,319 miles from each other. Without using pencil and paper, calculate how far apart the missiles are one minute before they meet.
3. A logician visits an island where two native tribes live. The members of one tribe will always tell the truth if asked a question. The members of the second tribe will always tell a lie if asked a question. The members of both tribes are identical.

   One day, the logician comes to a fork in the road on the island. One road from the fork leads to a village that the logician wishes to visit. The second road from the fork leads into the wilderness.

   A member of one tribe is standing at the roadside. The logician does not know if this tribesman is a truth teller or a lie teller.

   The logician thinks for a couple of minutes. Then he asks the tribesman one question. On hearing the answer, the logician now knows which road leads to the village.

   What was the question the logician asked?
4. I came across this little puzzle in a fantastic book of number puzzles by J. A. H. Hunter. A young lad went to a store to purchase a few items with a check. Asked how much he had spent, he replied, "If the check had shown half as many dollars and twice the number of cents, I'd have spent $2.58 less that I would have spent if the check had shown twice as many dollars and half as many cents. Can you figure out the value of the check?"
5. A man entered a store and bought a number of oranges at the rate of 3 for $1. He then bought a similar number of oranges in the store next door at a rate of 5 for $1. What was the average number of oranges bought for each dollar?

6. I found this perplexing little puzzle in Henry Ernest Dudeney's *Puzzles and Curious Problems*.

   Mr. Walker is going down a moving escalator and says that he counted 50 steps on the escalator as he did so.

   His friend, Mr. Trotman, said that he counted 75 steps but admitted that he was walking three times faster than Mr. Walker.

   If the escalator were stopped, how many steps would be visible? (We assume that each man moves at a uniform rate and that the speed of the escalator is also constant.)

7. A freight train is overtaken by a passenger train. The passenger train is traveling x times as fast as the freight train and takes x times as long to overtake the freight train as it takes to pass the freight train when both trains are going in opposite directions. What is the value of x?

8. Suppose you roll five six-sided dice. Which is more probable: that you will not roll a 6 or that you will roll *exactly* one 6?

9. Find all positive integers x and y that satisfy the following equation: $1 + x + x^2 = y^2$.

Solutions

1. Let x equal the five-digit number.

By putting a 1 at the front of the five-digit number, we have effectively created a number that is equal to 10^5 plus x. This number is three times smaller than if we had placed the 1 at the end of the number. We can therefore write:

$$10x + 1 = 3\,(100{,}000 + x)$$

$$10x + 1 = 300{,}000 + 3x$$

$$7x = 299{,}999$$

$$x = 42{,}857$$

Thus, our five-digit number is 42,857. The reader will find that $428{,}571 = 3 \times 142{,}857.$[1]

2. This question is easily answered if the solver goes the right way about it. The information that the missiles are 2,319 miles apart soon after the missiles are fired is superfluous and is really a red herring. Think of it this way. The *combined speed* of the two missiles is 30,000 miles per hour. Imagine a missile traveling at 30,000 miles per hour, or 500 miles per minute, going toward a stationary target. Now the answer is obvious. One minute before the missile hits the target, it is 500 miles from the target.

The same reasoning applies to the two missiles. One minute before the two missiles strike each other, they are 500 miles apart.[2]

3. There are two possibilities concerning the tribesman standing near the fork in the road. He is either a member of the truth-teller tribe or a member of the lie-teller tribe. There are also two possibilities concerning the road that the logician is pointing to. It is either the road that leads to the village or it is not.

The logician asks a question such that if the road he is pointing to leads to the village, the tribesman, regardless of his tribe, is forced to answer *yes* to the logician's question. If, however, the road the logician is pointing to does not lead to the village, that tribesman, regardless of his tribe, is forced to answer *no* to the logician's question.

Suppose the tribesman is a member of the truth-teller tribe.

The logician points to one fork of the road (let's assume that that road leads to the village, but the logician does not know that) and says, "*If* I were to ask you 'does this road lead to the village,' would you say yes?" If the road does lead to the village, the truth teller would answer *yes* to this question because if he was directly asked this question, he would answer *yes* to the question.

However, if the road does not lead to the village, the truthful tribesman would have to answer *no* to the logician's question because if the tribesman were directly asked if the road leads to the village, he will answer no. Therefore, the tribesman will answer truthfully (by saying no) that he would say no if asked if the road leads to the village.

Suppose, however, that the tribesman is a member of the lie-teller tribe. The logician points to one road (let's assume it is the road that leads to the village). The logician asks, "*If* I were to ask you 'does this road lead to the village,' would you say yes?" The untruthful tribesman is now forced to say *yes* to the logician's question because *if* the untruthful tribesman were asked directly if the road led to the village, he would lie and say *no*. But because of the way the logician's question is framed, he now has to lie and is forced to say *yes* in answer to the logician's question.

Suppose, however, that the road the logician points to is the road that does not lead to the village. The logician asks, "*If* I were to ask you 'does this road lead to the village,' would you say yes?" The untruthful tribesman is now forced to say *no* to the logician's question because *if* the untruthful tribesman were asked directly if the road led to the village, he would lie and say *yes*. But because of the way the logician's question is framed, he now has to lie and is forced to say *no* in answer to the logician's question.

Thus, if the tribesman answers yes (regardless of which tribe he belongs to) to the logician's question, the logician knows he is pointing to the road that leads to the village.

However, if the tribesman answers no (regardless of which tribe he belongs to) to the logician's question, the logician knows he is pointing to the road that does not lead to the village.[3]

4. Let the check's amount equal d dollars and c cents.
   From the question, we also know that

$$(\tfrac{1}{2}d + 2c) + 2.58 = 2d + \tfrac{1}{2}c \quad \text{Equation 1}$$

$$\text{Or } d + 4c + 5.16 = 4d + c \quad \text{Equation 2}$$

$$\text{Thus, } 3c + 5.16 = 3d \quad \text{Equation 3}$$

$$\text{Or } 1.72 = d - c \quad \text{Equation 4}$$

From the question, we know that the value of d must be even. Equation 3 tells us that d cannot equal 4 because if it did, the value of c would have to be more than 100, which is impossible. Therefore, the value of d must equal 2. Consequently, equation 4 tells us that the value of c must be 28. The value of the check was therefore 2.28 dollars.[4]

5. Many math enthusiasts answer this by saying that the average number of oranges bought for each dollar is 4. But this answer is incorrect.
   If the man had spent an equal number of dollars on the two kinds of oranges, then the average cost of the oranges would be 4 per $1, or $0.25 per orange. Thus, if he had spent, say, a dollar on each of the two kinds of oranges, he would have purchased a total of 8 oranges for $2. In that case, the average number of oranges purchased would be 4 per $1.

But in this little problem, it is the number of *oranges* that is equal, not the number of dollars.

We know that the man bought an equal number of oranges in each store. We can assume any number of oranges bought in each store, provided the number of oranges purchased in each store is identical. Let us suppose that the man purchased a total of 30 oranges in the two stores. This means he bought 15 oranges in the first store at the rate of 3 for $1. That purchase cost $5. In the second store, the man purchased 15 oranges at the rate of 5 for $1. That second purchase cost $3. The man has spent a total of $8 and has bought altogether 30 oranges. Therefore, the *average* number of oranges bought for $1 is 30/8, or 3.75.[5]

6. Let n equal the number of steps in the escalator. Let the time taken for one step to disappear at the bottom of the escalator be 1 unit of time.

Mr. Trotman takes 75 steps in $n - 75$ units of time, or 3 steps in $((n - 75)/25)$ units of time. Mr. Walker is moving three times slower than Mr. Trotman, so the number of steps Mr. Walker takes is 1 step in $((n - 75)/25)$ units of time. We also know from the question that Mr. Walker takes 50 steps in $n - 50$ units of time, or 1 step in $((n - 50)/50)$. This allows us to write

$$\frac{n - 75}{25} = \frac{n - 50}{50}$$

This equals

$$2n - 150 = n - 50$$

or

$$n = 100$$

Therefore, the total number of steps in the staircase is 100.[6]

7. This is a problem that may be solved by using a long, laborious, and tedious method.

However, if one recognizes that this is a problem in relative motion, it can be solved in the following straightforward manner.

Let the speed of the freight train equal 1. Let the combined length of the two trains also equal 1.

We are told that the speed of the passenger train is x times that of the freight train. Therefore, the speed of the passenger train is x.

The relative speed of the passenger train when overtaking is $x - 1$. When the two trains are traveling in opposite directions, the relative speed of the passenger train is $x + 1$. This allows us to write the following equation:

$$\frac{1}{x - 1} = \frac{x}{x + 1}$$

or

$$x^2 - x = x + 1$$

or

$$x^2 - 2x - 1 = 0$$

The positive value of x in this quadratic equation is $\sqrt{2} + 1$, or $2.4142135 +$. Thus, $x = 2.4142135 +$.[7]

8. The probability that you will not roll a 6 when you roll five dice is $(5/6)^5$, which equals $0.401877572 +$. The probability that you will roll *exactly* one 6 when you roll five dice is

$$\left(\frac{5}{6}\right)^4 \left(\frac{1}{6}\right)\left(\frac{5}{1}\right)$$

This works out as $0.482253086 \times 0.1666666 \times 5 = 0.401877572 +$. Therefore, the two probabilities are equal.

9. No integers satisfy the equation $1 + x + x^2 = y^2$.

Why? Consider the fact that x^2 is less than $(1 + x + x^2)$ and that $(1 + x + x^2)$ is less than $(x + 1)^2$. Therefore, $(1 + x + x^2)$ lies between two consecutive perfect squares. Consequently, it cannot be a perfect square also.

CHAPTER 22

Nine Curious and Beautiful Number Patterns

1.

The integers, beginning with 1, can be arranged to make this pretty pattern:

$$1 + 2 = 3$$

$$4 + 5 + 6 = 7 + 8$$

$$9 + 10 + 11 + 12 = 13 + 14 + 15$$

$$16 + 17 + 18 + 19 + 20 = 21 + 22 + 23 + 24$$

$$25 + 26 + 27 + 28 + 29 + 30 = 31 + 32 + 33 + 34 + 35 + 36$$

And so on.

2.

The formula for the nth triangular number is $(n(n + 1)/2)$. Thus, $T_1 = 1$, $T_2 = 3, T_3 = 6, T_4 = 10$, and so on.

The fact that triangular numbers elegantly fit in with the series of consecutive integers is clearly illustrated in the following pattern:

$$0 = 1T_0$$

$$1 + 2 + 3 = 2T_2$$

$$4 + 5 + 6 + 7 + 8 = 3T_4$$

$$9 + 10 + 11 + 12 + 13 + 14 + 15 = 4T_6$$

$$16 + 17 + 18 + 19 + 20 + 21 + 22 + 23 + 24 = 5T_8$$

And so on.

159

3.

The third pattern illustrates how consecutive square numbers are intrinsic to the mathematical structure of the series of integers. The number on the extreme far left of each row is equal to $(n(2n + 1))$, where n is the row number. In each row, the numbers immediately to the left and right of the equals sign are two sides of a *primitive Pythagorean triangle*:

$$3^2 + 4^2 = 5^2$$

$$10^2 + 11^2 + 12^2 = 13^2 + 14^2$$

$$21^2 + 22^2 + 23^2 + 24^2 = 25^2 + 26^2 + 27^2$$

$$36^2 + 37^2 + 38^2 + 39^2 + 40^2 = 41^2 + 42^2 + 43^2 + 44^2$$

And so on.

4.

Next, we see how the triangular numbers fit into the pattern involving the sum of squares:

$$1 + 2 = 3T_1$$

$$3^2 + 4^2 = 5^2$$

$$6 + 7 + 8 + 9 = 5T_3$$

$$10^2 + 11^2 + 12^2 = 13^2 + 14^2$$

$$15 + 16 + 17 + 18 + 19 + 20 = 7T_5$$

$$21^2 + 22^2 + 23^2 + 24^2 = 25^2 + 26^2 + 27^2$$

$$28 + 29 + 30 + 31 + 32 + 33 + 34 + 35 = 9T_7$$

$$36^2 + 37^2 + 38^2 + 39^2 + 40^2 = 41^2 + 42^2 + 43^2 + 44^2$$

and so on.

5.

The following figure illustrates the beautiful connections between the odd integers and cubic numbers:

$$1 = 1^3$$

$$3 + 5 = 2^3$$

$$7 + 9 + 11 = 3^3$$

$$13 + 15 + 17 + 19 = 4^3$$

$$21 + 23 + 25 + 27 + 29 = 5^3$$

$$31 + 33 + 35 + 37 + 39 + 41 = 6^3$$

And so on.

6.

The sixth patterns illustrates how the cubic numbers are connected to the integers in a surprising and beautiful way:

$$(0 + 1) \times 1 = 1^3$$

$$(1 + 2) \times 2 + 2 = 2^3$$

$$(1 + 2 + 3) \times 4 + 3 = 3^3$$

$$(1 + 2 + 3 + 4) \times 6 + 4 = 4^3$$

$$(1 + 2 + 3 + 4 + 5) \times 8 + 5 = 5^3$$

And so on.

7.

The seventh display is an unusual pattern, published here for the first time, involving the sums of triangular, square, and cubic numbers. The pattern neatly displays the surprising intimacy between these numbers:

Triangular	Square	Cubic	Sum
1	1	1	$(1 + 1 + 1)/3 = 1 = 1^2$
3	4	8	$(3 + 4 + 8)/3 = 5 = 1^2 + 2^2$
6	9	27	$(6 + 9 + 27)/3 = 14 = 1^2 + 2^2 + 3^2$
10	16	64	$(10 + 16 + 64)/3 = 30 = 1^2 + 2^2 + 3^2 + 4^2$
15	25	125	$(15 + 25 + 125)/3 = 55 = 1^2 + 2^2 + 3^2 + 4^2 + 5^2$

And so on.

8.

The eighth display of the integers that I give here must rank as one of the most beautiful patterns in mathematics. Many students of recreational mathematics are stunned when they first encounter it:

$$(1^3) = 1^2$$

$$(1^3 + 2^3) = (1 + 2)^2$$

$$(1^3 + 2^3 + 3^3) = (1 + 2 + 3)^2$$

$$(1^3 + 2^3 + 3^3 + 4^3) = (1 + 2 + 3 + 4)^2$$

$$(1^3 + 2^3 + 3^3 + 4^3 + 5^3) = (1 + 2 + 3 + 4 + 5)^2$$

And so on.

9.

The ninth display is a beautiful pattern involving successive triangular numbers, beginning with 1:

$$1 + 3 + 6 = 10$$

$$15 + 21 + 28 + 36 = 45 + 55$$

$$66 + 78 + 91 + 105 + 120 = 136 + 153 + 171$$

$$190 + 210 + 231 + 253 + 276 + 300 = 325 + 351 + 378 + 406$$

And so on.

I will close this chapter by mentioning that the great Hungarian mathematician Paul Erdős (1913–1996) was once asked, "Why are numbers beautiful?"

Erdős responded, "It's like asking why is Ludwig van Beethoven's Ninth Symphony beautiful. If you don't see why, someone can't tell you. I know numbers are beautiful. If they aren't beautiful, nothing is."

CHAPTER 23

Six Little-Known Results in Number Theory

1.

One astonishing result in number theory is the beautiful theorem that the French mathematician Adrien Marie Legendre (1752–1833) discovered in 1808. He discovered that any natural number, n, is equal to the number of 1s in its binary representation plus the number of 2s in the prime factorization of n!

For example, let n equal 4. Now, consider 4 factorial. It is usually written by the mathematician as 4! It equals $4 \times 3 \times 2 \times 1$, or 24. The prime factorization of 24 is $2^3 \times 3$. Therefore, the number of 2s in the prime factorization of 4! is 3. In binary notation, 4 is written as 100. Thus, the number of 1s in the binary notation of 4 is 1. Add those two numbers, 3 and 1, together, and one obtains 4, which is the number we commenced with. This is an example of Legendre's beautiful theorem.

Here is a second example. Let n equal 5. Now consider 5!, which equals 120. The prime factorization of 120 equals $2^3 \times 3 \times 5$. Therefore, the number of 2s in the prime factorization of 5! is 3. In binary notation, 5 is written as 101. Therefore, the number of 1s in the binary representation of 5 is 2. Adding 3 and 2 together gives 5, and we started with 5!

2.

We will now consider the number of odd coefficients in any row, n, of *Pascal's triangle*. Let us first write a few rows of Pascal's triangle. Row 0 begins with 1 at the apex and assumes that there is a 0 at either end of the 1. (The first row is called row 0 so that various properties of the triangle are easily explained.) Each succeeding row is numbered row 1, row 2, row 3, and so on. Each number in each row is the sum of the two numbers immediately above it. The first seven rows of Pascal's triangle are shown below:

1

1 1

1 2 1

1 3 3 1

1 4 6 4 1

1 5 10 10 5 1

1 6 15 20 15 6 1

1 7 21 35 35 21 7 1

It can easily be proved that if and only if n is a prime number, then every number (excluding the two 1s at either end of each row) in row n of Pascal's triangle is divisible by n. For example, consider row 2. We know that 2 is prime, so the numbers in row n (there is only one number in row n other than the two 1s) is divisible by 2. Row 3 contains two 3s. Because 3 is a prime number, every number in row 3 other than the two 1s is divisible by 3.

Consider row 4. It contains three composite numbers. We find that not every number in row 4 is divisible by 4. That is because 4 is a composite number.

Consider row 5. It contains two prime numbers and two composite numbers. All four numbers are divisible by 5 because 5 is a prime.

This property of Pascal's triangle is remarkable in itself. But there is more.

Consider any number, n.

If n is a prime, let g equal the number of 1s in its binary representation. It turns out that the number of coefficients that are odd numbers in row n will always be a power of 2. But these are not any old powers of 2. In fact, for all n, the number of coefficients in row n that are *odd* numbers will equal 2^g. This is a totally unexpected result. This property of Pascal's triangle always holds. (This is a special case of Lucas's theorem. The theorem was proposed in 1878 by the French mathematician Édouard Lucas.)

We will look at some examples to clarify the matter.

Let n equal 3. Express 3 in binary representation. It is 11. Thus, there are two 1s in the binary representation for 3. Therefore, g equals 2. Consider now the number of odd coefficients in row 3 of Pascal's triangle. The number of odd coefficients in row 3 is 2^2. (They are 1, 3, 3, 1.) Here, g equals 2, because there are two 1s in the binary representation of 3.

Consider the integer 5. The binary representation for 5 is 101. There are two 1s here. Therefore, g equals 2. Thus, the number of odd coefficients in row 5 of Pascal's triangle is, once again, 2^2, or 4. We find that row 5 is 1, 5, 10, 10, 5, 1.

SIX LITTLE-KNOWN RESULTS IN NUMBER THEORY   167

Consider row 7. In binary representation, 7 is 111. Thus, in this case, g equals 3. Therefore, the number of odd coefficients in row 7 is 2^3, or 8. We find that this is indeed correct. The odd coefficients in row 7 are 1, 7, 21, 35, 35, 21, 7, 1.

Readers familiar with Pascal's triangle will know that the sum of every row is a power of 2.

At this point, the reader might be tempted to believe that there is nothing remarkable about all this. She may believe that all coefficients in row n, where n is a prime number, will be odd coefficients. But that is not the case. In fact, we can say this: all the coefficients in row n of Pascal's triangle are odd if, and only if, n equals $2^n - 1$.

Consider row 11. The coefficients in row 11 are 1, 11, 55, 165, 330, 462, 462, 330, 165, 55, 11, 1. When we express 11 in binary representation, we obtain 1011. There are three 1s in its binary representation. Thus, g equals 3. Therefore, the number of odd coefficients in row 11 of Pascal's triangle is 2^3, or 8. We find that this is indeed the case. The odd coefficients are 1, 11, 55, 165, 165, 55, 11, 1.

3.

Suppose you wish to find the number of numbers that are relatively prime (i.e., numbers that have no common factor) to, say, 489. The following is a little-known method that always works. First, find the prime factors of 489. They are 3 and 163. Now decrease each prime factor by 1. This gives you 2 and 162. Multiply 2 by 162. The answer is 324. This tells us that there are 324 positive integers less than 489 that are relatively prime to 489.

Here is a second example. How many positive integers are relatively prime to 521? Because 521 is a prime, simply subtract 1 from it, obtaining 520. Thus, there are 520 integers less than 521 that are relatively prime to 521.

Here is a third example. How many positive integers are relatively prime to 3,080? The prime factors of 3,080 are 2^3, 5, 7, and 11. Decrease each prime factor by 1, obtaining the four numbers 2^2, 4, 6, and 10. The product of these four numbers is 960. Therefore, the number of positive integers less than 3,080 that are relatively prime to 3,080 is 960.

4.

In 1959, the Hungarian mathematical genius Paul Erdős (1913–1996) was introduced to a Hungarian child prodigy named Lajos Pósa (1947–). Erdős had been told that the 11-year-old was exceptionally good at mathematics. Over lunch, Erdős gave the boy the following problem: prove that if one chooses $2n$ positive integers and one selects $n + 1$ integers from the list, two of those numbers must be relatively prime.[1]

The young boy thought about the problem for a few moments. Then he said that two of the numbers must be consecutive. Therefore, they must be relatively prime.

The young boy was right, of course. Erdős was extremely impressed by the young boy's intellectual ability. Two consecutive integers must be relatively prime. Suppose one chooses the numbers from 1 to 8. Here, n equals 4. One could pick 2, 4, 6, and 8. These are all multiples of 2 and so are not relatively prime to any other. But if one picks ($n + 1$), or five integers from the eight numbers, one's choice must include two consecutive integers, and these are always relatively prime. Erdős had actually thought about the problem that he posed to the boy years earlier, and at that time, it took him 10 minutes to solve.

Lajos Pósa went on to win first place in the International Mathematical Olympiad in 1966 and came in second place in 1965. Pósa went on to become a very famous mathematician in Hungary.[2]

To see why two consecutive numbers must be relatively prime, let the two consecutive numbers be n and $n - 1$. Assume that they are not relatively prime. That means an integer, d, that is greater than 1 divides both n and $n - 1$. If that is the case, then d divides $n - (n - 1)$. But $n - (n - 1) = 1$. An integer greater than 1 cannot divide evenly into 1. This contradiction proves that two consecutive integers must be relatively prime.

5.

Suppose you are given a positive integer, n. Is there a way to find a number that has n divisors that include the number itself and 1 among the divisors?

Yes, there is. Say that n equals 10. In other words, you want to find a number that has 10 divisors. Find the prime factors of 10. They are 2 and 5. Subtract 1 from each prime factor to obtain 1 and 4. Place these numbers in descending order: 4, 1. Raise successive primes, beginning with 2, to these powers and multiply the results: $2^4 \times 3^1 = 48$. This tells us that 48 has 10 positive factors. These are 1, 2, 3, 4, 6, 8, 12, 16, 24, and 48.

Here is a second example. Suppose you want to find a positive number with 60 positive divisors. The prime factors of 60 are 2, 2, 3, 5. Subtract 1 from each factor, obtaining 1, 1, 2, 4. Arrange the resulting numbers in descending order. One obtains 4, 2, 1, 1. Raise successive primes to these powers and multiply the results: $2^4 \times 3^2 \times 5^1 \times 7^1 = 5,040$. Thus, the number you seek is 5,040.[3] This number has more divisors than any smaller number. (Note that this method does not guarantee that the number you seek is the smallest such number. For example, using this method, suppose one wishes to find a number with eight divisors. The method described here gives 30 as the solution That is correct; 30 does have eight divisors. But 24, which is smaller than 30, also has eight divisors.)

6.

Suppose you wish to find the number of divisors of a positive number, n. Let's say that n equals 24. Proceed as follows. Find the prime factors of 24. They are 2^3 and 3^1. Add 1 to each exponent and multiply the results: $(3 + 1) \times (1 + 1) = 8$. Therefore, there are eight divisors of 24. That's all there is to it. The eight divisors of 24 are 1, 2, 3, 4, 6, 8, 12, and 24.

Let's say that n equals 735. The prime factors of 735 are 3, 5, and 7^2. Add 1 to each exponent. This gives 2, 2, 3. Multiply those three numbers together. The answer is 12. Therefore, the number 735 has 12 positive divisors. These are 1, 3, 5, 7, 15, 21, 35, 49, 105, 147, 245, and 735.

CHAPTER 24

Twenty-Six Ridiculous Questions

The following questions are asked for fun and should not be taken too seriously. No advanced knowledge of mathematics or logic is required to solve any of them. I was asked these problems during my childhood from various people.

1. Suppose you are a runner in a race. You are positioned down the field. You then increase your speed, and you overtake the runner who is in second position. What position are you now in?

2. A teenager purchased a shirt for $97. He borrowed $50 from his mom and $50 from his dad. Soon after he purchased the shirt, he returned $1 to his mom and $1 to his dad and kept $1 for himself. He now finds that he owes $49 + $49, or $98, plus the $1 he kept for himself. This comes to $99. Where did the other $1 go?

3. Three men can build a wall in 20 days. How long would it take 2 men to build the same wall?

4. A man can climb a stairs from the first floor to the third floor in 10 minutes. How long will it take him to climb the stairs from the third floor to the sixth floor?

5. In an extremely busy city, you are the driver of a bus. At the first bus stop, 10 people get off and 12 people get on. At the second bus stop, 7 people get off and 11 people get on. At the third bus stop, 8 people get off and 10 people get on. How old is the bus driver?

6. There is a large lily growing in the center of a circular pond. Each day, the area covered by the lily is doubled. At the end of 20 days, the pond is completely covered by the lily. When was the pond half covered by the lily?

7. A snail can climb out of a 20-foot deep well at the rate of 3 feet every day. However, at night, the snail slips back down 2 feet. How long will it take the snail to climb out of the well?

8. Two vagabonds sit down on the side of the road to have supper. One vagabond has 5 loaves of bread, and the other has 3 loaves. A third, female vagabond, who has $8 in her pocket, joins them. The 3 vagabonds share the 8 loaves equally between them. The third vagabond hands over her $8. How should the $8 be shared by the two male vagabonds?

9. If an airplane crashes on the U.S.–Canada border, where are the survivors buried?

10. A man looks at a photograph and states, "Brothers and sisters I have none, but that man's father is my father's son." Who is in the photograph?

11. A circular swimming pool that has a radius of 20 feet costs $50,000. How much does a swimming pool with a radius of 40 feet cost?

12. If it is one-fifth the time from now to midnight as it is from now to midday, what time is it now?

13. How much earth is in a hole that measures 1 foot in length, 1 foot in width, and 1 foot in depth?

14. How many times in 24 hours do the hour hand and the minute hand overlap?

15. What word is pronounced differently in New York and Boston?

16. What is 10 divided by 1/2 plus 30 divided by 1/3?

17. If your doctor gives you three pills and advises you to take one every 30 minutes, how long will the pills last?

18. Mary's mother has three daughters. One is called Ann. The second is called Betty. What is the name of the third daughter?

19. What is the name of the ship in the movie *Mutiny on the Bounty*?

20. Aunt Mary was in her living room, reading the newspaper, with a very serious look on her face. "Look at this, George," she said to her husband. "It states here that the rate of undetected murders in this country has dramatically increased in the last three years. It says police sources have indicated that they expect this trend to continue well into the future This is very concerning. Where will it all end?"

    "I have never heard so much nonsense in my life," George said.

    Should Aunt Mary have been concerned with the newspaper story?

21. What word is spelled incorrectly in every dictionary?

22. What is the first word in Spain's national anthem?

23. Suppose a deck of 52 cards is thoroughly shuffled and 5 cards are then dealt from the deck. Is it more probable that the hand of 5 cards will contain exactly 2 red cards or exactly 3 red cards?

24. A New York retired army sergeant was recently walking through Manhattan when he saw an old army buddy from Los Angeles whom he had not seen for 30 years and with whom he had lost all contact.

    They got talking. "My daughter is going to college at present," the Los Angeles buddy said.

    "That's interesting! And what is her name?" asked the New Yorker.

    "Mary Ann."

    "How nice! The same name as her mother's."

    How did the New Yorker know that?

25. A triangular lot for sale in a residential part of town was recently advertised. The proposed sale of the lot attracted a lot of interest from potential buyers. The advertisement stated that the three sides of the triangular plot measured 38 by 43 by 82 yards. But one potential buyer said there must be a mistake in the advertisement. Why did he believe this?

26. What does the following product equal: $1 \times 2 \times 3 \times 4 \times 5 \times 6 \times 7 \times 8 \times 9 \times 0$?

Solutions

1. Second position. Since you overtook the runner in second position, you are now in second position.[1]
2. There is no missing dollar. The teenager's calculations are incorrect. He borrowed $100 from his parents. He paid $97 for a shirt. That left him with $3. He repaid $1 to his mom and $1 to his dad and kept $1 for himself. That accounts for the $100 that he borrowed. The teenager ended up owing $98, not $99.[2]
3. 30 days. Two men can do 2/3 of the work done by 3 men. Therefore, the time it takes 2 men to complete the job is 3/2 × 20, or 30 days.[3]
4. 15 minutes. From the first floor to the third floor is two stories. From the third floor to the sixth is three stories. Since he can climb each story in 5 minutes, it will take him 5 minutes longer to go from the third floor to the sixth floor.[4]
5. The answer is your present age. You are the driver of the bus![5]
6. 19 days.[6]
7. 18 days. At the end of the 17th day, the snail has climbed 17 feet. He then climbs 3 more feet on the 18th day, and so he is out of the well.[7]
8. Each vagabond ate 2 and 2/3 loaves of bread. Therefore, the first male vagabond contributed 1/3 of a loaf of bread to the communal meal. The second male vagabond contributed 2 and 1/3 loaves to the meal. These contributions are in a 7-to-1 ratio. Therefore, the $8 should be divided among the two male vagabonds in a 7-to-1 ratio. The first vagabond should get $1, and the second vagabond should get $7.[8]
9. The *survivors* of the airplane accident would not be buried![9]
10. His son. The man has no brothers. Therefore, when he states, "but that man's father is my father's son," he must be referring to himself. He is therefore stating that the father of the man in the photograph is himself. Therefore, the man in the photograph is his son.[10]
11. $200,000. The area of the second swimming pool is four times larger than the first.
12. 10:00 p.m.[11]
13. None. There is no earth inside a hole![12]
14. The hour hand and the minute hand of a clock overlap 22 times in a 24-hour period.[13]
15. Differently.
16. 20 + 90, or 110.
17. One hour. You take one pill immediately, the second pill after 30 minutes, and the third pill after 60 minutes.[14]
18. Mary.[15]
19. *Bounty.*
20. No, she should not have been concerned. As George recognized, the rate of *undetected* murders in any society is unknown![16]
21. Incorrectly.[17]
22. Spain's national anthem contains no words.

23. The two probabilities are identical. If the 5 cards dealt contain exactly 2 red cards, it must contain 3 black cards. Therefore, the probability that 2 red cards are dealt is exactly the same as the probability of dealing 3 black cards. That in turn means the probability of dealing 2 red cards is identical to the probability of dealing 3 red cards.

24. The New York sergeant was male. His old army buddy from Los Angeles was female. Her name—and her daughter's name—was Mary Ann.[18]

25. The potential buyer of the lot knew that the sum of any two sides of a triangle must exceed the length of the third side. The dimensions given in the advertisement for the triangular lot violated this basic rule of triangles. Therefore, the potential buyer knew that a triangular lot with the dimensions given is not possible.

26. Zero.

Notes

Chapter 1

1. "Al Capone: Biography," Britannica, https://www.britannica.com/biography/Al-Capone, accessed February 25, 2022.

2. "Seemingly Amazing Coincidences May Not Be So Unlikely after All," *Daily Mail*, April 2, 2011, https://www.dailymail.co.uk/home/moslive/article-1371572/Seemingly-amazing-coincidences-unlikely-all.html, accessed February 25, 2022.

3. "Remembering John Parr and George Ellison," British Legion, https://www.britishlegion.org.uk/stories/the-first-and-last-soldiers-to-be-killed-in-wwi, accessed February 25, 2022.

4. "South Africa's Lottery Probed as 5, 6, 7, 8, 9, and 10 Drawn and 20 Win," BBC, July 8, 2014, https://www.bbc.com/news/world-africa-55154525, accessed February 11, 2022.

5. "Five People Who Were in the Wrong Place, at the Wrong Time," Mental Floss, May 8 2022, https://www.mentalfloss.com/article/57582/5-people-who-were-wrong-place-wrong-time-multiple-times, accessed February 25, 2022.

6. "Laura Buxton Balloon Coincidence," Snopes, https://www.snopes.com/fact-check/whether-balloon/, accessed February 25, 2022.

7. "Inside John Wilkes Booth's Famous Family," History.com, October 27, 2020, https://www.history.com/news/john-wilkes-booth-family, accessed February 25, 2022.

8. Leila Schneps and Coralie Colmez, *Math on Trial: How Numbers Get Used and Abused in the Courtroom* (New York: Basic Books, 2013), chapter 1.

9. Ray Hill, "Reflections on the Cot Deaths Cases." *Medicine, Science, and the Law* 47, no. 1 (June 2016): 2–6.

10. "Google Breaks Record for Calculating Digits of *Pi*," Sky News, May 14, 2019, https://news.sky.com/story/google-breaks-record-for-calculating-digits-of-pi-11665198, accessed May 8, 2022.

Chapter 2

1. Viktor T. Toth, "Why Is Hawking Evaporation Directly Proportionate to Mass?" Quora, March 20, 2017, https://www.quora.com/Why-is-Hawking-radiation-inversely-proportional-to-mass, accessed May 8, 2022.

2. Isaac Asimov, *A Choice of Catastrophes: The Disasters That Threaten Our World* (New York: Fawcett Columbine Books, 1979), 153.

3. "The Heat Experienced at a Campfire Is Proportional to the . . ." Socratic.org, November 28, 2016, https://socratic.org/questions/the-heat-experienced-by-a-hiker-at-a-campfire-is-proportional-to-the-amount-of-w, accessed February 25, 2022.

4. "Marine Propulsion," *Physics Forums*, December 22, 2015, https://www.physicsforums.com/threads/marine-propulsion.849313/, accessed February 25, 2022.

5. "Wind Power," *Energy Education*, https://energyeducation.ca/encyclopedia/Wind_power, accessed May 8, 2022.

6. "Orbits and Kepler's Laws. Kepler's Third Law: The Squares of the Orbital Periods of the Planets Are Directly Proportional to the Cubes of the Semi-Major Axes of Their Orbits," NASA, https://solarsystem.nasa.gov/resources/310/orbits-and-keplers-laws/#:~:text=Kepler's%20Third%20Law%3A%20the%20squares,the%20radius%20of%20its%20orbit, accessed February 25, 2022.

7. On-Line Encyclopedia of Integer Sequences, March 31, 2015, https://oeis.org/A021578, accessed February 25, 2022.

8. "Mathematicians Find a Completely New Way to Write the Number 3," *New Scientist*, September 18, 2019, https://www.newscientist.com/article/2216941, accessed February 25, 2022.

9. James D. Harper, "Ramanujan, Quadratic Forms, and the Sum of Three Cubes," *Mathematics Magazine* 86, no. 4 (2013): 275–79, doi:10.4169/math.mag.86.4.275.

10. "A Question on Ryley's Theorem: Any N as the Sum of Three Rational Cubes," Narkive, https://sci.math.narkive.com/aWY4EjqW/a-question-on-ryley-s-theorem-any-n-as-the-sum-of-three-rational-cubes, accessed May 8, 2022.

11. "Find Three Consecutive Positive Integers a, b, and c Such That $a + b$ Is a Perfect Square and $b + c$ Is a Perfect Cube," Math Stack Exchange, August 21, 2016, https://math.stackexchange.com/questions/1898908/find-three-consecutive-positive-integers-a-b-and-c-such-that-ab-is-a-perfect-sq, accessed February 25, 2022.

12. Henry Ernest Dudeney, *536 Puzzles and Curious Problems*, ed. Martin Gardner (New York: Charles Scribner's Sons, 1967), 36–37.

13. "Question 67.B. Find All Possible Sets of Three Consecutive Integers Whose Cubes Add Up to a Perfect Square," *Mathematical Gazette* 67, no. 441 (October 1983): 228–30.

14. "Cubic Number," Wolfram Math World, https://mathworld.wolfram.com/CubicNumber.html, accessed May 8, 2022.

15. "Cubic Number."

Chapter 3

1. "Numerators of Continued Fraction Convergents to the Square Root of 2," On-Line Encyclopedia of Integer Sequences, https://oeis.org/AOO1333, accessed February 25, 2022.

2. *American Mathematical Monthly* 63, no. 4 (April 1956): 247. Gustave Robson credits this form of the proof to Robert James Gauntt.

3. Ronald Sprague, *Recreation in Mathematics* (London: Blackie & Son Ltd., 1963), 37–38.

Chapter 4

1. Jeffrey M. Kubina and Marvin C. Wunderlich, "Extending Waring's Conjecture to 471,600,000," *Mathematics of Computation* 55, no. 192 (1990): 815–20.

2. "Cubic Number," November 4, 2022, Wolfram MathWorld, https://mathworld.wolfram.com/CubicNumber.html, accessed February 25, 2022.

3. "Numbers Whose Shortest Representation as a Sum of Positive Cubes Requires Exactly 8 Cubes," On-Line Encyclopedia of Integer Sequences, December 5, 2018, https://oeis.org/AO18889, accessed May 8, 2022.

4. "Waring's Problem," November 4, 2022, Wolfram MathWorld, https://mathworld.wolfram.com/WaringsProblem.html, accessed February 25, 2022.

Chapter 5

1. Minoru Tanaka, "Numerical Investigation on Cumulative Sum of the Liouville Function," *Tokyo Journal of Mathematics* 3, no. 1 (June 1980): 187–89.

2. "Project Euler 46: Odd Number Not a Prime plus Twice a Square," Math Blog, https://www.mathblog.dk, accessed February 25, 2022.

3. "Prime Number Races," University of British Columbia, May 6, 2010, https://personal.math.ubc.ca/~gerg/slides/Hanover-6May10b.pdf, accessed May 8, 2022.

4. "608,981,813,029," ProofWiki, https://proofwiki.org/wiki/608,981,813,029, accessed May 8, 2022.

5. "Littlewood and Number Theory," Indian Academy of Sciences, September 2013, https://www.ias.ac.in/describe/article/reso/018/09/0789-0798, accessed February 25, 2022.

6. "Quotations, Leonhard Euler," MacTutor, https://mathshistory.st-andrews.ac.uk/Biographies/Euler/quotations/, accessed February 25, 2022.

Chapter 6

1. "James Clerk Maxwell: A Force for Physics," Physics World, December 1, 2006, https://physicsworld.com/a/james-clerk-maxwell-a-force-for-physics/, accessed February 25, 2022.

2. "Fermat's Last Theorem. Definition, Example, and Facts," Encyclopedia Britannica, https://www.britannica.com/science/Fermats-last-theorem, accessed February 25, 2022.

3. Tamar Friedmann and C. R. Hagen, "Quantum Mechanical Derivation of the Wallis Formula for Pi," *Journal of Mathematical Physics* 56: 112–201.

4. "Why Can't an Energy Level Exist Containing 0.9 Electron Wave?" Stack Exchange, June 8, 2017, https://physics.stackexchange.com/questions/338068/why-cant-an-energy-level-exist-containing-0-9-electron-wave-wavelengths-why-mu, accessed February 25, 2022.

5. "What Do Imaginary Numbers Practically Represent in the Schrödinger Equation?" Stack Exchange, December 20, 2017, https://physics.stackexchange.com/questions/375363/what-do-imaginary-numbers-practically-represent-in-the-schr%C3%B6dinger-equation, accessed February 25, 2022.

6. Bryan Deaton, "The Dirac Equation and the Positron," Duke Physics, https://webhome.phy.duke.edu/~kolena/modern/deaton.html, accessed February 26, 2022.

7. "Higgs Boson," Wikipedia, https://en.wikipedia.org/wiki/Higgs_boson, accessed May 8, 2021.

8. "Introduction to Quantum Mechanics," Wikipedia, https://en.wikipedia.org/wiki/Introduction_to_quantum_mechanics, accessed February 26, 2022.

9. "The Borsuk-Ulam Theorem," Princeton, https://www.cs.princeton.edu/courses/archive/spr05/cos598B/lec13.pdf, accessed September 13, 2021.

10. "NASA Solar System Exploration," NASA, https://solarsystem.nasa.gov/planets/neptune/in-depth/, accessed May 9, 2021.

Chapter 8

1. "Assassination of Robert F. Kennedy," Wikipedia, https://en.wikipedia.org/wiki/Assassination_of_Robert_F._Kennedy, accessed January 26, 2022.

2. "Anton Horner Dies; Played Solo Horn for 28 Years," *New York Times*, December 7, 1971, https://www.nytimes.com/1971/12/07/archives/anton-horner-dies-played-solo-horn.html, accessed January 26, 2022.

3. Roger Highfield, "The Name Game—The Weird Science of Nominative Determinism," *Evening Standard*, April 10, 2012, https://www.standard.co.uk/lifestyle/the-name-game-the-weird-science-of-nominative-determinism-6384728.html, accessed March 7, 2022.

4. "Ann and Frank Webb—The B.T.S. Community Board," British Tarantula Society, February 23, 2004, https://thebts.co.uk/forums/forum/b-t-s-tarantula-community-board/bts-discussion-forum/203-ann-and-frank-webb, accessed March 7, 2022.

5. "Keith Weed Appointed as the New RHS President," Gardenforum, July 31, 2020, https://www.gardenforum.co.uk/news/people/keith-weed-appointed-as-the-new-rhs-president/, accessed March 7, 2022.

6. "Fiona Lander," Business News, https://www.businessnews.com.au/Person/Fiona-Lander, accessed March 7, 2022.

Chapter 10

1. "Quark," Encyclopedia Britannica, http://www.britannica.com/EBchecked/topic/486323/quark, accessed January 10, 2023.

2. "Wave–Particle Duality," Wikipedia, https://en.wikipedia.org/wiki/Wave%E2%80%93particle_duality, accessed January 10, 2023.

3. Roger Penrose, "Is Mathematics Invented or Discovered?" April 13, 2020, https://www.google.com/search?client=firefox-b-1-d&q=Roger+Penrose%2C+-+%E2%80%9CIs+Mathematics+Invented+or+Discovered#fpstate=ive&vld=cid:83470474,vid:ujvS2K06dg4, accessed August 17, 2020.

4. Ying Zhang, "Representing Primes as $x^2 + 5y^2$: An Inductive Proof That Euler Missed," ARXIV, Cornell University, October 3, 2006, https://arxiv.org/abs/math/060654, accessed May 5, 2021.

5. Janos Pintz, "On Legendre's Prime Number Formula," *American Mathematical Monthly* 87 (1980): 733–35.

6. "Hiroshima, Nagasaki, and Subsequent Weapons Testing," World Nuclear Association, March 2016, https://www.world-nuclear.org/information-library/safety-and-security/non-proliferation/hiroshima,-nagasaki,-and-subsequent-weapons-testin.aspx, accessed January 10, 2023.

7. To name just three: Sean Caroll (1966–), physicist, California Institute of Technology; Alan Guth (1947–), physicist, Massachusetts Institute of Technology; and Andrei Linde (1948–), physicist, Stanford University.

Chapter 12

1. "Frequency of Easter Sundays," Webspace, https://webspace.science.uu.nl/~gent0113/easter/easter_text2b.htm, accessed January 11, 2023.

2. Elwyn R. Berlekamp, John H. Conway, and Richard K. Guy, *Winning Ways for Your Mathematical Plays* (Boca Raton, FL: CRC Press, 2001).

Chapter 14

1. Jonathan Mayo, *The 1966 World Cup Final: Minute by Minute* (London: Short Books Ltd., 2016).

2. "FIFA World Cup Final. Minute by Minute. Average and Total Attendance 1930–2018," Statista, https://www.statista.com/statistics/264441/number-of-spectators-at-football-world-cups-since-1930/, accessed May 8, 2022.

Chapter 15

1. Clara Moskowitz, "The 11 Most Beautiful Mathematical Equations," LiveScience, March 22, 2022, https://www.livescience.com/57849-greatest-mathematical-equations.html, accessed February 25, 2022.

Chapter 17

1. "Euclid-Mullin Sequence," Wikipedia, https://en.wikipedia.org/wiki/Euclid%E2%80%93Mullin_sequence, accessed February 25, 2022.

Chapter 18

1. "Phases of the Moon: 1901 to 2000," AstroPixels, http://astropixels.com/ephemeris/phasescat/phases1901.html, accessed January 12, 2023; "Moon Phases April 1912," http://www.calendar-12.com/moon, accessed January 12, 2003.

2. "Titanic," Wikipedia, https://en.wikipedia.org/wiki/Titanic, accessed February 25, 2002; "Eastern Air Lines Flight 401," Wikipedia, https://en.wikipedia.org/wiki/Eastern Air Lines Flight 401, accessed February 25, 2022.

3. "Edward Smith (Sea Captain)," Wikipedia, https://en.wikipedia.org/wiki/Edward_Smith_(sea_captain), accessed February 25, 2022.

4. "Eastern Airlines Flight 401," Wikipedia, https://en.wikipedia.org/wiki/Eastern_Air_Lines_Flight_401, accessed February 25, 2022. The Captain of Flight 401 was Robert Albin Loft.

5. Thomas Turnerbegan, *The Band That Played On: The Extraordinary Story of the Eight Musicians Who Went Down with the Titanic* (Nashville, TN: Thomas Nelson, 2011).

6. "Flight 401 Down in the Glades," Flamingo, February 26, 2018, https://flamingomag.com/2018/02/26/flight-401, accessed February 25, 2022.

7. "Discover the Ship," Titanic.com, http://www.titanic.com/Discover Ship, accessed February 25, 2022.

Chapter 19

1. Patrick Lyon, "Captain of the Titanic Captured on Camera Peering through a Telescope," *Mirror*, October 20, 2016, https://www.mirror.co.uk/news/uk-news/captain-titanic-captured-camera-peering-9086428, accessed February 25, 2022.

Chapter 21

1. Martin Gardner, *The Magic Numbers of Dr. Matrix* (Amherst, NY: Prometheus Books, 1985). A similar puzzle to the one given here is found in chapter 2 (pp. 30, 245, 247, and 248).

2. Martin Gardner, *The Colossal Book of Short Puzzles and Problems*, ed. Dana Richards (New York: W. W. Norton, 2006), Problem 4.11, "Colliding Missiles," pp. 87 and 94.

3. Gardner, *The Colossal Book of Short Puzzles and Problems*, Problem 13.22, "Fork in the Road," pp. 373, 383–86.

4. J. A. H. Hunter, *Fun with Figures* (New York: Dover Publications, 1965), Puzzle 54, pp. 33 and 104.

5. Gilbert Wilkinson, *The Complete Home Entertainer: Games and Amusements for the Whole Family* (London: Odhams Press, 1940), Brain Twister and Torturers, Puzzle 3, "Oranges for Owlglass," pp. 190 and 481.

6. Henry Ernest Dudeney, *Puzzles and Curious Problems* (London: Thomas Nelson and Sons, 1932), Puzzle 68, "The Moving Staircase," pp. 27 and 135.

7. L. A. Graham, *The Surprise Attack in Mathematical Problems* (New York: Dover Publications, 1968), "Passing Trains," pp. 56 and 57.

Chapter 23

1. Bruce Schechter, *My Brain Is Open: The Mathematical Journeys of Paul Erdős* (New York: Simon & Schuster, 2000).

2. "Lajos Pósa (Mathematician)," Wikipedia, https://en.wikipedia.org/wiki/Lajos_P%C3%B3sa_(mathematician), accessed February 25, 2022.

3. Anonymous, "What Is the Smallest Number That Has 8 Divisors?" Quora, January 11, 2019, https://www.quora.com/What-is-the-smallest-number-that-has-8-divisors, accessed February 25, 2022.

Chapter 24

1. "Passing Second in a Race," Riddle Brain Teasers, https://riddlesbrainteasers.com/passing-second-in-a-race, accessed February 25, 2022.

2. "I Saw a Shirt for $97. I Borrowed $50 from Mom and $50 from Dad," Quora, March 9, 2019, https://www.quora.com/I-saw-a-shirt-for-97-I-borrowed-50-from-mom-and-50-from-dad-After-purchasing-I-return-1-to-mom-and-1-to-dad-and-kept-1-for-me-I-now-owe-49-49-98-plus-1-I-kept-my-self-which-is-99-Where-did-1-go, accessed February 25, 2022.

3. "If It Takes 3 Men to Build a Wall in 10 Days, How Long Will It Take 2 Men to Build the Same Wall?" Quora, https://www.quora.com/If-it-takes-3-men-to-build-a-wall-in-10-days-how-long-will-it-take-2-men-to-build-the-same-wall, accessed May 8, 2022.

4. "It Takes You 50 Seconds to Walk from the 1st Floor to the 3rd," Algebra.com, https://www.algebra.com/algebra/homework/word/misc/Miscellaneous_Word_Problems.faq.question.216771.html, accessed February 25, 2022.

5. "You're a Bus Driver. At the First Stop 4 People Get On," Youth4work.com, https://www.youth4work.com/Talent/Common-Sense/Forum/123093-youre-a-bus-driver-at-the-first-stop-4-people-get-on-at-the-second, accessed February 25, 2022.

6. Gilbert Wilkinson, *The Complete Home Entertainer* (London: Odhams Press, 1940), 195, 196, 483.

7. "A Snail Climbs Out of a Well 3 Feet during the Day, and during Night, It Slips 2 Feet. If the Well Is 30 Feet Deep, How Long Will It Take for the Snail to Climb?" Quora, https://www.quora.com/A-snail-climbs-a-well-3-feet-during-the-day-and-during-night-it-slips-2-feet-If-the-well-is-30-feet-deep-how-long-will-it-take-for-the-snail-to-climb, accessed May 8, 2022.

8. Maurice Kraitchik, *Mathematical Recreations* (New York: Dover Publications, 1953), "The Problem of the Pandects," p. 28.

9. "A Plane Crashed on the Border of U.S. and Canada. Where Do They Bury the Survivors?" Brainzilla.com, https://www.brainzilla.com/brain-teasers/riddles/kVe3Boy7/a-plane-crashed-on-the-border-or-us-and-canada-where-do-they-bury-the, accessed February 25, 2022.

10. Wilkinson, *The Complete Home Entertainer*, 121, 122, 474. This is a very similar problem to the one posed here.

11. "Half as Long till Midnight," Braingle.com, https://www.braingle.com/brainteasers/1245/half-as-long-till-midnight.html. This similar puzzle to the one presented here, accessed May 8, 2022.

12. "How Much Dirt Is There in a Hole 3 Feet Deep, 6 Ft Long and 4 Ft Wide?" Quora, https://www.quora.com/How-much-dirt-is-there-in-a-hole-3-feet-deep-6-ft-long-and-4-ft-wide, accessed February 25, 2022.

13. "How Many Times a Day Do a Clock's Hands Overlap?" Quora, https://www.quora.com/How-many-times-a-day-do-a-clock%E2%80%99s-hands-overlap-1, accessed February 25, 2022.

14. "A Doctor Gives You Three Pills and Tells You to Take One Every Half an Hour. How Long Do the Pills Last?" Brainzilla.com, https://www.brainzilla.com/brain-teasers/riddles/eO2jqaOd/a-doctor-gives-you-three-pills-and-tells-you-to-take-one-every-half-an, accessed February 25, 2022.

15. "Johnny's Mother Had Three Children. The First Child Was Named April. The Second Child Was Named May. What Was the Third Child's Name?" Quora, https://www.quora.com/Johnny-s-mother-had-three-children-The-first-child-was-named-April-The-second-child-was-named-May-What-was-the-third-child-s-name. This is a similar puzzle to the one given here, accessed February 25, 2022.

16. Wilkinson, *The Complete Home Entertainer*, 120, 121, 474.

17. "Which Word in the Dictionary Is Spelled Incorrectly?" Brainzilla.com, https://www.brainzilla.com/brain-teasers/riddles/q8y0q9OD/which-word-in-the-dictionary-is-spelled-incorrectly, accessed February 25, 2022.

18. Wilkinson, *The Complete Home Entertainer*, 118, 119, 474.

References for Further Reading

Chapter 1

Hand, David. *The Improbability Principle: Why Coincidences, Miracles, and Rare Events Happen Every Day*. London: Transworld Publishers, 2014.

Mazur, Joseph. *Fluke: The Maths and Myths of Coincidences*. London: Oneworld Publications, 2016.

Weaver, Warren. *Lady Luck: The Theory of Probability*. New York: Dover, 1982.

Chapter 2

Simmons, Gustavus J. "Palindromic Powers." *Journal of Recreational Mathematics* 3, no. 2 (1970): 93–98.

Wells, David. *The Penguin Dictionary of Curious and Interesting Numbers*. New York: Penguin, 1988.

Chapter 3

Flannery, David. *The Square Root of 2: A Dialogue Concerning a Number and a Sequence*. New York: Copernicus Books, 2006.

Gardner, Martin. *"The Square Root of 2."* In *A Gardner's Workout*. Natick, MA: A. K. Peters, 2002.

O'Shea, Owen. "The Square Root of 2." In *The Call of the Primes: Surprising Patterns, Peculiar Puzzles, and Other Marvels of Mathematics*. Amherst, NY: Prometheus Books, 2016.

Chapter 4

Gardner, Martin. "Waring's Problems." In *Knotted Doughnuts and Other Mathematical Entertainments*. New York: W. H. Freeman, 1986.

Roberts, Joe. "Waring's Problem." In *Lure of the Integers*. Washington, DC: Mathematical Association of America, 1992.

Chapter 6

Balaguer, Mark. *Platonism and Anti-Platonism in Mathematics*. Reprint ed. New York: Oxford University Press, 1998.

Panza, Marco, and Andrea Sereni. *Plato's Problem: An Introduction to Mathematical Platonism.* London: Palgrave Macmillan, 2013.

Chapter 7
Gardner, Martin. "Elegant Triangles." In *Mathematical Circus.* London: Pelican Books, 1981.

Chapter 9
Gardner, Martin. "The Calculus of Finite Differences." *In Martin Gardner's New Mathematical Diversions.* Chicago: University of Chicago Press. 1966.

Chapter 10
Carroll, Sean. *Something Deeply Hidden: Quantum Worlds and the Emergence of Spacetime.* New York: Dutton, 2019.
Gardner, Martin. "Physics: End of the Road." In *Gardner's Whys and Wherefores.* Chicago: University of Chicago Press, 1989.
Goldschmidt, Tyron. *The Puzzle of Existence: Why Is There Something Rather Than Nothing?* London: Routledge, 2013.
Greene, Brian. *The Hidden Reality: Parallel Universes and the Deep Laws of the Cosmos.* New York: Vintage Books, 2011.
Kolakowski, Leszek. *Why Is There Something Rather Than Nothing: Questions from Great Philosophers.* London: Penguin, 2008.
Rovelli, Carlo. *Reality Is Not What It Seems: The Journey to Quantum Gravity.* New York: Riverhead Books, 2018.

Chapter 11
Gardner, Martin, and Silvanus P. Thompson. *Calculus Made Easy.* New York: St. Martin's Press, 1998. (The appendix in this book is titled "Some Recreational Problems Related to Calculus.")
Nahin, Paul J. *When Least Is Best: How Mathematicians Discovered Many Clever Ways to Make Things as Small (or as Large) as Possible.* Princeton, NJ: Princeton University Press, 2021.

Chapter 14
Gardner, Martin. *The Magic Numbers of Dr. Matrix.* Amherst, NY: Prometheus Books, 1985.
O'Shea, Owen. *The Magic Numbers of the Professor.* Washington, DC: Mathematical Association of America, 2007.

Chapter 15
Garrett, Pat. *The Authentic Life of Billy the Kid.* Norman: University of Oklahoma Press, 1954.
Guinn, Jeff. *The Last Gunfight: The Real Story of the Shootout at the OK Corral and How It Changed the American West.* London: The Robson Press, 2012.
Stiles, T. J. *Jesse James: Last Rebel of the Civil War.* New York: Vintage Books, 2003.

Chapter 20
Langley, Philippa, and Michael Jones. *The King's Grave: The Search for Richard III.* New York: St. Martin's Press, 2013.